ACS SYMPOSIUM SERIES **545**

Polymeric Drugs and Drug Administration

Raphael M. Ottenbrite, EDITOR
Virginia Commonwealth University

Developed from a symposium sponsored
by the Division of Polymer Chemistry, Inc.,
at the 204th National Meeting
of the American Chemical Society,
Washington, D.C.,
August 23–28, 1992

American Chemical Society, Washington, DC 1994

Library of Congress Cataloging-in-Publication Data

Polymeric drugs and drug administration / Raphael M. Ottenbrite [editor].

p.   cm.—(ACS symposium series, ISSN 0097–6156; 545)

"Developed from a symposium sponsored by the Division of Polymer Chemistry, Inc., at the 204th National Meeting of the American Chemical Society, Washington, DC, August 23–28, 1992."

Includes bibliographical references and indexes.

ISBN 0–8412–2744–6

1. Polymeric drugs—Congresses.   2. Polymeric drug delivery systems—Congresses.

I. Ottenbrite, Raphael M. II. American Chemical Society. Division of Polymer Chemistry. III. American Chemical Society. Meeting (204th: 1992: Washington, D.C.) IV. Series.

RS201.P65P642   1994
615'.7—dc20   93–48086
                                                        CIP

The paper used in this publication meets the minimum requirements of American National Standard for Information Sciences—Permanence of Paper for Printed Library Materials, ANSI Z39.48–1984. ∞

PRINTED IN THE UNITED STATES OF AMERICA

Foreword

THE ACS SYMPOSIUM SERIES was first published in 1974 to provide a mechanism for publishing symposia quickly in book form. The purpose of this series is to publish comprehensive books developed from symposia, which are usually "snapshots in time" of the current research being done on a topic, plus some review material on the topic. For this reason, it is necessary that the papers be published as quickly as possible.

Before a symposium-based book is put under contract, the proposed table of contents is reviewed for appropriateness to the topic and for comprehensiveness of the collection. Some papers are excluded at this point, and others are added to round out the scope of the volume. In addition, a draft of each paper is peer-reviewed prior to final acceptance or rejection. This anonymous review process is supervised by the organizer(s) of the symposium, who become the editor(s) of the book. The authors then revise their papers according to the recommendations of both the reviewers and the editors, prepare camera-ready copy, and submit the final papers to the editors, who check that all necessary revisions have been made.

As a rule, only original research papers and original review papers are included in the volumes. Verbatim reproductions of previously published papers are not accepted.

M. Joan Comstock
Series Editor

Contents

INDEXES

Preface

CONTROLLED DRUG DELIVERY is a well-established, multidisciplinary area of science and technology. New advanced polymeric systems allow one to administer drugs to the target site at the desired concentration. The therapeutic level of the drug at the dysfunction site can be sustained over a prolonged period of time, resulting in enhanced drug efficacy and improved patient compliance. This technique minimizes potential drug toxicity, protects other organs against possible side effects, and improves treatment safety in general.

Polymeric materials are an intrinsic part of drug delivery systems designed for controlled drug release. The versatility of polymer applications in medicine is reflected in this book. The chapters touch upon the use of polymer-based systems in oncology, for treatment of inflammatory processes and diabetes, for prevention of gastric ulcers, and for management of central nervous system conditions.

From the standpoint of the properties of polymeric materials, this book provides discussions on bioelastic characteristics, micelle formation, hydrogels, and drug release based on the electric properties of polymers. Polymer degradation via different routes such as dissolution, erosion, enzyme degradation, and photodegradation makes these materials particularly suitable for drug delivery devices. Theoretical aspects for designing drug delivery systems are represented by computer simulations of interactions between polymer matrices and solvents.

In summary, this volume gives a broad view of the latest advances in medicinal applications of polymers. This book should be of interest to specialists involved in this exciting area of research.

I greatly appreciate the efforts of my assistant, Natasha Fadeeva. Her understanding of the field and her organizational skills were invaluable throughout the production of this book.

RAPHAEL M. OTTENBRITE
Department of Chemistry
Virginia Commonwealth University
Richmond, VA 23284–2006

RECEIVED August 23, 1993

Chapter 1

Polymer Systems for Biomedical Applications
An Overview

Raphael M. Ottenbrite and Natalya Fadeeva

Department of Chemistry, Virginia Commonwealth University,
Richmond, VA 23284-2006

During the past two decades, bioapplications have become one of the largest areas for polymer use, with an annual consumption growth of 6% *(1)*. Conventional polymers, such as poly(vinyl chloride), polystyrene, polyethylene, and polypropylene, are being replaced with high performance materials that provide new and more effective ways of administering drugs and enhance therapeutic efficacy. New advanced polymers used as drugs and in delivery systems allow one to diminish side effects and toxicity of drugs, and to improve patient compliance.

Polymers can exert biological activity directly as polymeric drugs, such as polysulfonates, polycarboxylates, and other polyanionic polymers. Polymers also provide the means for controlled delivery of low molecular weight drugs in the form of polymeric drug carriers or polymeric prodrugs. Polymeric drug delivery systems allow for a variety of routes for drug administration (oral, parenteral, transdermal, nasal, ocular, etc.). In cases when the activity of conventional drugs is lost or diminished in the body environment, pharmacokinetics and pharmacodynamics of the active agent can be modified by combining drugs with macromolecules. This technique can enhance the efficacy of conventional and newly engineered pharmaceuticals, alleviates immunologic response from the host, and reduces biological inactivation of the therapeutic agent. Under certain conditions, site-specific drug delivery and sustained drug release can provide a therapeutic environment at the site of dysfunction over prolonged periods of time.

Polymer-based systems intended for medicinal applications are designed with consideration of a number of factors such as: the nature of disease, drug properties, type of therapy (acute or chronic), physiology of the patient, administration route, therapy site, and characteristics of the polymeric material used, including the mechanism of drug delivery or/and sustained release (drug diffusion through membranes or from monolithic devices, or based on osmotic pumps). Polymeric materials, presently utilized in medicine, include synthetic polymers that mimic natural substances, naturally occurring macromolecules, and chemically modified natural polymers. These modifications are made to improve biocompatibility, degradability, or to introduce other desired properties.

0097–6156/94/0545–0001$08.00/0

Proteins and Polypeptides in Drug Delivery

Presently, proteins and polypeptides are being extensively investigated for controlled drug delivery and sustained drug release. For example, elastomeric polypeptides with bioelastic characteristics present interesting and promising systems for pharmaceutical applications. Bioelastic properties are related to the ability of proteins to hydrophobically fold and aggregate in response to an increase in ambient temperature. Collagen and elastin are two of many natural bioelastic materials used for biomedical applications. Collagen is a structural protein which renders strength to the skin and bones. When collagen is heated, its helicoidal structure is destroyed with the formation of randomly organized gelatin. Formaldehyde treated collagen produces a material that is used in artificial heart valves, while cross-linked gelatin is used to improve surface blood-compatibility of diaphragms in artificial heart devices (2). Elastin, an amorphous elastic protein, has been used in the development of artificial connective tissues (2).

Bioelastic materials with a broad range of mechanical properties, from rigid to flexible, can be prepared by copolymerization of amino acids. Synthetic polypeptides, mimicking natural compounds, can be designed to possess a certain degree of biodegradability. The fact that the products of these polymers degradation are mostly amino acids allows for a good biocompatibility. For example, cross-linked matrices of poly(Gly-Val-Gly-Val-Pro), obtained by gamma-irradiation of poly(GVGVP) solutions (3,4), have been shown to undergo contraction/relaxation cycles (Chapter 1). These cycles occur as a deswelling/swelling process in water with a ten-fold volume increase when temperatures is lowered from 37°C to 20°C. Macromolecular matrices can be loaded with therapeutic agents by swelling in an aqueous drug solution followed by contracting to expel the excess of water on raising the temperature. Swell-loading is particularly beneficial when used with hydrophobical drugs that have limited solubility in water, such as steroids, peptides, oligonucleotides, or small proteins.

To enhance diffusional release of the agent from drug-laden matrices by degradation, the bioelastic polymers can be synthesized to include sites in the backbone which are susceptible to proteolytic enzymes. "Chemical clocks" have been developed, such as carboxamides of asparagine or glutamine residues. Depending on the residue location in the sequence, the carboxamide will hydrolytically cleave at the matrix-milieu interface with a predictable half-life (3). The introduction of an occasional ester in the backbone, by replacing a glycine residue with a glycolic acid residue, allows for chain cleavage at the matrix interface. The carboxylates formation enhances swelling. The matrix composition can be altered to change affinities for various pharmaceuticals (5).

The medicinal activity of some drugs is based on DNA interaction (Chapter 2). However, low specificity and selectivity, as well as severe toxic effects have been associated with these drugs. Sequence specific DNA-binding agents have been developed which are used as chemotherapeutic agents, structural probes for DNA, and artificial restriction enzymes (6,7). One approach is to design agents which are selective to certain DNA sequences based on the recognition of parts of human genome rich in guanine-cytosine (GC) or adenine-thymine (AT) sequences *(8)*. Oligopeptides netropsin and distamycin, which display antibiotic and antiviral properties, are known to bind specifically to 4 and 5 contiguous AT base pairs in the minor groove of double helical ß-DNA and thus block the template function of DNA *(6,8,9)*. The substitution of a heterocycle alters the strict AT binding preference and permits GC recognition *(10,11)*. The development of drugs that recognize long GC rich sequences is based on the design of optimal oligoimidazole-carboxamido analogs of distamycin as the "reading frame" for the longer agents linked with a flexible polyethylene tether. The linkers must meet the structural criteria which include suitable flexibility, shape, and lipophilicity *(12-14)*. The ligands can also serve as "vectors" directing the delivery of other agents to the targeted DNA sequences.

Protein-based systems are used to provide glucose-sensitive release of insulin for the treatment of diabetes. Insulin-releasing systems include encapsulation of viable pancreatic cells, electromechanical systems consisting of a glucose sensor and an insulin injection pump, and various chemical systems. One type of chemical systems is based on a signal transduction *(15)* which uses glucose-sensitive enzymes and transduces the glucose signal to other physicochemical signals, such as pH or redox. An example of a pH sensitive system is a porous cellulose membrane covered with immobilized poly(acrylic acid) and glucose oxidase (GOD) *(16)*. In the presence of glucose, GOD catalyzes the conversion of glucose to gluconic acid, lowering the pH and inducing conformational changes in the graft polymer chain. In turn, this alters the size of the membrane pores *(17,18)*. The redox sensitive system is represented by insulin immobilized on a composite membrane through a disulfide linkage *(19,20)*. In a new protein device (Chapter 3), GOD is directly coupled to insulin via S-S bonds. The disulfide bonds can be reversibly cleaved and insulin released in response to the presence of glucose molecules.

Polysaccharides in Drug Delivery

Polysaccharides are another group of biopolymers effectively used in drug delivery. Natural polysaccharides can be classified according to their origin (algal, botanical, microbial, and animal) *(21)*. Structurally, these materials can be homo- or heteropolysaccharides. Homopolysaccharides are formed from one type of monosaccharide. They include starch, cellulose, dextrans, inulin, and chitin. Heteropolysaccharides are formed from two or more monosaccharide types, and a majority of polysaccharides belong to this group. The fact that polysaccharides can form physical gels under different conditions and are biodegradable makes them particularly suitable for using in drug delivery systems.

With respect to water solubility, polysaccharides vary from insoluble (chitin, cellulose) to swellable in water (agarose, furcelleran, dextrans, starch) to water soluble (gum arabic, xanthan, pullulan). Hydrogels prepared from natural polysaccharides are basically characterized by adequate biocompatibility and are susceptible to enzyme degradation. Starch *(22, 23)* and dextran *(24)* are among the polysaccharides commonly used as biodegradable hydrogels. Hydrogels can be loaded with high molecular weight drugs, such as proteins or polypeptides, by swelling dry hydrogel material in the drug solution. However, drug loading may be restricted by the low penetration of the excipient such as proteins into the hydrogel. Hydrogel preparation by polymerization of monomers or by crosslinking water-soluble polymer molecules often involves the use of external crosslinkers which may impair the biological activity of the proteinoid drugs being incorporated. One method to obviate the use of undesirable cross-linking agents and to maintain the biological activity of the therapeutic agent, is to introduce double bonds into the polymers with the subsequent cross-linking under gamma-irradiation *(25-27)*. Thus, drug-loaded hydrogels can be prepared from dextran without adding a crosslinking agent (Chapter 4). After a high molecular weight enzyme (invertase) is incorporated into dextran, the dextran matrix is functionalized by introducing double bonds and then crosslinked under gamma-irradiation. It has been shown that gamma-irradiation does not impair the biological activity of the incorporated invertase. The hydrogels formed do not require further purification due to the absence of external crosslinking agents in the system.

Cyclodextrins have been shown to significantly improve the therapeutic efficacy of several anticancer agents. Cyclodextrins are composed of 6 to 8 glucose molecules which are linked by α-1,4-bonds. The structure is cone-like or donut-like, with a hydrophilic exterior and a hydrophobic interior. Non-polar molecules can be solubilized using cyclodextrins if the size of non-polar portion of the molecule is comparable to that of the cyclodextrin lipophilic cavity. The formation of "inclusion complexes" between cyclodextrin (host) and hydrophobic drug (guest) takes place without bond formation. Therefore, the complexes are reversible and in dynamic equilibrium with the free drug in solution.

In cancer boron-neutron capture therapy, para-boronophenylalanine (BPA) is used to deliver [10]boron to malignant melanoma cells. Poor solubility of p-boronophenylalanine at physiological pH does not allow for adequate therapeutic levels in the cells. The solubility is improved by forming complexes between p-boronophenylalanine and cyclodextrins (Chapter 5). A mixture of 2-hydroxypropyl-β-cyclodextrins, has been shown to be non-toxic even after repeated parenteral administration *(28-30)*. The cyclodextrins-BPA complexes increase the delivery of BPA-mediated boron into the blood stream 15-fold compared to the BPA-buffer formulations.

Control Release Systems Based on Electric Signal Stimulation

Sustained drug release from the polymer matrix can be triggered by interaction with a solvent, changes in the ambient conditions (pH, temperature), enzymatic degradation, ultrasonic irradiation, and application of electromagnetic fields. The latter involves polymers with intrinsic electric conductivity. Conductive polymeric materials are currently being used in the manufacture of lightweight secondary batteries, hybrid composite materials, and electroconductive coatings. In the context of drug delivery devices, drug release from a conductive polymeric matrix under electrochemical pulse stimulation presents a valuable method of controlled drug release *(31-33)*.

One group of polymers used for electrically induced drug release are pyrrole polymers and copolymers which are electrically conductive in the oxidized form. The advantage of polypyrroles compared to other groups of conducting polymers is their stability in the conductive state *(34)*. Polypyrrole films are obtained by electropolymerization of pyrrole from inorganic electrolyte solutions. Anionic drugs can be loaded into cationic polypyrroles and then released from the conductive polymeric matrix in response to the electrical signal. Monitoring the potential, current, or charge allows one to control the amount of ionic drug released from the matrix.

Cationic drugs can also be used for loading polypyrrole films after the modification of polypyrrole ion dynamics. Polypyrroles, modified with different ionic compounds, can form composite polymeric materials with cation-exchange properties. These materials can bind or release cationic drugs under the influence of electrical field applied (Chapter 6). A variety of neuroleptic drugs, such as phenothiazine derivatives and tricyclic antidepressants, can be released from the polypyrrole matrix in response to electrochemical pulse *(35-37)*.

Electrochemical reactions, carried out in aqueous solutions, generate hydronium ions (H^+) at the anode, while hydroxyl ions (OH^-) are produced at the cathode. This process changes the local pH near the electrodes. These pH changes can be utilized to effect controlled drug release in an "on-off" mode due to ion exchange *(39)*. Under electric current stimulation, "on-off" surface erosion has been observed for specific polymeric complexes *(40,41)*. The polymer surface erosion is caused by disruption of hydrogen bonds in the complex due to ionization of proton donor, or by deionization of the polymer complex pair based on the locally induced pH changes. Rapid "on-off" response to an electric current signal can effect ion exchange, deswelling at the anode, and electric osmosis from the anode to the cathode (depending on the electric field strength). An "on-off" release of a positively charged solute from negatively charged hydrogels is facilitated by hydronium ions generated at the anode under an electric current. In Chapter 7, the release of neutral solutes, such as hydrocortisone, which are physically entrapped inside a gel, is described. The process is based on passive diffusion and gel contraction caused by deswelling at the anode side.

Synthetic Hydrogels in Drug Delivery

In contrast to reservoir systems with zero order kinetics release, diffusion control for monolithic matrix systems is based on a square root time relationship. Hydrophilic synthetic polymers, including poly(ethylene oxide) (PEO) based systems, have been proposed as monolithic drug delivery devices *(42-49)*. In an aqueous environment, the polymer swells and the active ingredient diffuses out of the device. The polymeric matrix swells at a constant rate as long as the penetrant concentration at the moving boundary remains constant. To maintain zero order release from a monolithic device, a number of requirements have to be met, such as: constant surface area, constant swelling rate of the polymer matrix, and high diffusivity of the loaded agent. Release of an active agent requires a balance between the diffusion of the active agent and the solubilization of the polymer matrix (Chapter 8). Drug diffusivity, polymer matrix swelling, and polymer solubilization rate can be modified by changing molecular weight of the polymer, or by blending polymers with different molecular weights.

Mucoadhesion is another important characteristic of polymer matrices for monolithic drug delivery devices, particularly for oral administration. To a great extent, mucoadhesion depends on viscoelastic properties of the polymer gel. To prevent the swollen outer layer from flowing, elastic characteristics should dominate over the viscous properties. This allows the polymer gel to stay in place for a longer time with the maximum level of mucoadhesion. To avoid the displacement of the mucoadhesive device due to tissue movements, the polymer gel should also meet shear module requirements.

Due to the flexible structure of PEO macromolecules, this polymer forms entangled physical bonds which can deeply interpenetrate into mucous membrane of gastrointestinal tract. The water-swelling, mucoadhesive, and viscoelastic properties of PEO gels are functions of the polymer molecular weight and are more pronounced for higher molecular weight materials *(50,51)*. For example, PEO with Mw 20,000 has no bioadhesive properties, while PEO with Mw 4,000,000 PEO displays a very good bioadhesion *(52)*. By blending different molecular weights PEO fractions, it is possible to modify water swelling behavior and mucoadhesive properties of PEO systems, and thus affect the drug release characteristics of polymeric drug delivery systems (Chapter 8).

Polymeric Micelles as Drug Carriers

The therapeutic efficacy of many low molecular weight drugs can be improved by combining them with polymeric carriers. Polymeric micelles, composed of AB-type block-copolymers, contain a hydrophobic core which facilitates the incorporation of hydrophobic drugs *(53)*. After conjugation with a polymer hydrophilic outer shell, hydrophobic drugs often tend to precipitate due to a high localized drug concentration. However, a core-shell structure of micelles inhibits intermicellar aggregation of the hydrophobic cores and thus maintains an adequate water solubility which allows one to achieve a proper structure for drug targeting. By incorporating hydrophobic drugs into the micelle core, the drug is being protected against inactivation by enzymes or other bioactive species *(54)*. The drug release is controlled by stability of micelles, the core hydrophobicity, and the chemical nature of the bond between the drug and the polymer.

In polymeric micelles, the drug carrier functions are shared by structural segments of block-copolymer. Drug delivery *in vivo* is controlled by the outer shell which is responsible for interactions with the bioenvironment and determines the pharmacokinetic behavior and biodistribution of the drug, while the inner core expresses pharmacological activity. This heterogenous structure is favorable for the formation of a highly functionalized carrier system compared to conventional polymeric carrier systems. On the other hand, polymeric micelles are based on intermolecular noncovalent interactions and are in equilibrium with a single polymer chains. Polymer chains with a molecular weight lower than critical values for renal filtration are described in Chapter 9. These polymers can be excreted by the renal route without causing toxic effects associated with long-term drug accumulation.

Interactions between Polymers and Cells

Polymers, such as synthetic polyanionic electrolytes, can express bioactivity directly, as polymeric drugs. In the early 1960's, pyran (copolymer of divinyl ether and maleic anhydride) was found to possess anticancer activity *(55)*. Since then several synthetic polycarboxylic acid polymers have been evaluated for biological activities, such as stimulation of the reticuloendothelial system, modulation of humoral and/or cell mediated immune responses, and improvement of resistance to various microbial infections *(56-60)*.

Polyanions are cytotoxic against both DNA and RNA pathogenic viruses from several major virus groups with diverse characteristics. The advantage of synthetic polyanions is the ability to protect mammalian hosts against a broad variety of viruses, compared to the narrow antiviral spectrum of most conventional drugs used for chemotherapy. Synthetic polyanions can also provide prolonged protection against viral infection *(61,62)*. Antiviral activity of polyanions is affected by the molecular weight and the structure of the polymer. The molecular weight of polymeric drug should not exceed 50,000 to allow for excretion from the host to prevent long term toxicity *(63)*. On the other hand, low molecular weight polyanions (<30,000) were shown to be ineffective against viral infections.

The anti-HIV activity of a polyanionic polymer, poly(maleic acid-alt-2-cyclohexyl-1,3-dioxepin-5-ene) [poly(MA-CDA)], is described in Chapter 10 which deals with three major issues, such as polymer evaluation, optimization of the direct synthesis of poly(MA-CDA), and increasing molecular weight of [poly(MA-CDA)] by polymer-polymer grafting.

To allow for the interaction between polyanionic polymers and target cells, the polymer has to attach to the cell membrane with the subsequent cellular uptake. However, interactions between cell membranes and polyanionic polymers are hindered by electric repulsion between the negatively charged cell surface and the anionic groups on the polymers. One method to improve the affinity of these polymers for cell membranes is to encapsulate the polymer in mannan coated liposomes, which results in significant increase in antitumor activity as compared to the control *(64)*. This effect could be due to specific interactions between the sugar moieties and receptors on the macrophage. The nonspecific interactions between polyanionic polymers and cell membranes can be enhanced by grafting hydrophobic groups onto the polymer *(65)*. This technique was applied to the poly(MA-CDA) system, using negatively charged liposomes as a model for evaluating the membrane affinity of the polymer (Chapter 11). The following mechanism was suggested for the biological activity of modified poly(MA-CDA). First, the polymer attachment to the cell membrane occurs, facilitated by the improved membrane affinity. Second, the attached polymer is being incorporated into cells where it stimulates the signal transduction system. This stage may be affected by the molecular weight of poly(MA-CDA). The *in vitro* tests on cultured cell lines indicate that lower molecular weight polymers elicit higher activities. Third, the signal is transduced to NADPH-oxidase located on the membrane and activates the enzyme to produce superoxide. The polymers are cytotoxic at relatively high concentrations, and the cytotoxicity correlates with the improved membrane affinity.

The mechanism of cell interactions with high molecular weight polymers has also been studied from the standpoint of polymeric prodrugs, where the polymer is the carrier. The behavior of prodrugs or drug conjugates in the body is determined by physicochemical properties of the prodrug as well as by the biological environment *(66,67)*. Normal cells are permeable to polymers depending on the polymer molecular weight, with a drastic change around Mw 30,000. However, the tissue inflammation produces enhanced membrane permeability with the subsequent leakage of proteins and lipids from the blood vessels into the interstitial space *(68)*.

To investigate the relationship between the polymer properties (including its molecular weight) and the polymer distribution between the normal and inflammatory sites, radiolabelled poly(vinyl alcohol) (PVA) was used (Chapter 12). The inflammation was induced in mice by injecting carrageenan. It was found that PVA accumulated at the inflammatory sites at higher rates and higher concentrations than at the normal sites. The increased molecular weight facilitated the PVA accumulation at the inflammatory sites, with the maximum accumulation for the Mw 200,000 PVA. However, the life-time for higher Mw PVA in the blood circulation was longer than that for the lower Mw polymer. Thus, the profile of the PVA accumulation at the inflammatory site depends on the balance between the rate of PVA accumulation at the inflammatory site and the period of PVA retention in the blood, which may be achieved at some intermediate value of PVA molecular weight.

Hydrolytically Degradable Polymers in Medicine

Polymer biodegradation is a broad concept which includes various mechanisms, such as photodegradation, hydrolysis, enzymatic degradation and thermo-oxidative degradation. In the process of being degraded in an aqueous environment, such as the human body, the polymeric material undergoes hydration, loss of strength, loss of mass integrity, and loss of mass. Ideally, polymer degradation would occur with the formation of natural metabolites and the biodegradable system would leave no residual material in the body after the device has performed its function *(69)*. Biodegradable polymers have been investigated for a wide variety of biomedical application, such as suture materials, drug delivery systems, and ligature clamps.

Hydrolysis is one of the most common degradation mechanisms. Polymers undergo hydrolytical degradation due to the presence of hydrolytically unstable bonds, hydrophilic enough to allow for the water access. One group of hydrolytically degradable polymers are aliphatic polyesters where degradation occurs via hydrolysis of the ester linkages in the polymer backbone *(70)*. Toxicity related to drug delivery devices is commonly caused by the products of polymer degradation rather than by the polymer itself. Therefore, it is important for the degradation products to be harmless to the host at the concentrations being present in the body.

Poly(lactic acid) (PLA) *(71)*, polyparadioxane *(72)*, and a copolymer of glycolic acid and trimethylene carbonate *(73)* are examples of commercial products that meet the above requirements. Upon degradation, PLA produces lactic acid (a natural metabolite), while the other two polymers yield harmless degradation products.

Polymeric microspheres can serve as carriers of a therapeutic agent. The release of the drug incorporated into microspheres is controlled by diffusion of the drug from the microspheres. The diffusion process is influenced by several factors, such as structure, porosity, and surface morphology of microspheres. Hydrolytical decomposition of microspheres results in increased porosity and enhances water access to the drug. In turn, it facilitates drug release based on drug dissolution and diffusion. Chapter 11 presents the results of research aimed at developing polymeric microspheres for drug delivery. Copolymers of 1,5-dioxepan-2-one and D,L-dilactide and of 1,3-dioxan-2-one and ε-caprolactone were synthesized, and the *in vitro* degradation was studied using coordination type initiators. Along with poly(adipic anhydride) (PAA) and poly(lactide-co-glycolide) (PLG), these copolymers were used to prepare microspheres for drug delivery. The copolymers of 1,5-dioxepan-2-one and D,L-dilactide are amorphous and degrade in an uniform way with the formation of water soluble products. The degradation time for oligomers varies over a broad range spanning from years for DXO-rich copolymers to 30-40 days with the D,L-LA rich copolymers.

With regard to microsphere morphology, the SEM studies indicate that the surface roughness and porosity depends both on the polymer concentration in the microspheres and on the polymer selected. The PLG microspheres with the highest polymer concentration have a very smooth and nonporous surface. As the concentration of polymer is decreased, the surface becomes uneven, accompanied with enhanced porosity. Microspheres obtained from the crystalline PAA, however, have a rough surface, porous walls and a hollow center. By blending these two polymers in different ratios, microspheres with different density, porosity and surface morphology can be obtained.

Polyesterpeptides, a class of alternating polyesteramides of α-hydroxy acids and α-amino acids *(74-77)*, have been shown to be subject to hydrolysis and biodegradation both in *in vitro (78)* and *in vivo (79,80)*. Degradable polyesteramide materials for potential biomedical and other applications, containing no α-hydroxy or α-amino acid moieties, are described in Chapter 12. First type, PEA-I, was prepared by a two-step polycondensation reaction from 1,6-hexanediol, adipoyl chloride and 1,6-hexanediamine, while the second type, PEA-II, was synthesized via the anionic copolymerization of ε-caprolactam and ε-caprolactone. Based on selective ester hydrolysis by base treatment, it has been shown that both types of polyesteramides have random chain structures. The hydrolysis of these polyesteramides in buffer solutions is facilitated by increasing the temperature and ester content of the copolymers. Hydrolysis of the copolymers can be performed under both acid and base conditions.

Drug-Polymers Conjugation

In many cases, the use of drugs with a high therapeutic potential is restricted due to severe side effects. Systemic drug toxicity can be significantly reduced by the site-directed delivery of the drug which limits the activity to the location of the disease. The chemical conjugation of drugs to macromolecules presents an efficient way of modifying pharmacological and biochemical characteristics of the drug.

Misoprostol, a synthetic 16-hydroxy analog of natural prostaglandin E_1, is used for its gastric antisecretory and mucosal protective properties. Misoprostol prevent gastric ulcer formation in people who are being treated with non-steroidal anti-inflammatory preparations. The optimum therapeutic affect is observed when Misoprostol is applied directly to the gastric mucosa. Side-effects of Misoprostol include uterotonic activity if the drug gets into the circulation, and diarrhea in case of intestinal exposure. In order to reduce these side-effects without impairing the drug therapeutic efficacy, it is desirable to provide delivery of misoprostol only to the stomach. A number of functionalized polymer-based systems, designed to slowly release misoprostol in the stomach but not in the intestines, have been developed *(81)*. The drug release mechanism is based on a covalent silicon ether bond to the C-11 hydroxy group of misoprostol. The silyl linker is cleaved under acidic conditions (pH 1-3) with the release of intact misoprostol from the polymer matrix, while at the higher pH (>5), such as in the intestine, the silyl bond is stable. Controlled delivery of Misoprostol to the stomach provides adequate therapeutic levels at the treatment site and minimizes systemic and intestinal exposure to the drug. Extended periods of the drug availability in the stomach due to controlled release from the delivery system also allow one to reduce the dosing frequency (Chapter 15).

Another example of a potent therapeutic agent with undesirable side-effects is 5-fluorouracil (5FU) which has a remarkable antitumor activity. The side effects of 5-FU have been significantly diminished by the conjugation of the drug with the polysaccharides α-1,4-polygalactosamine (PGA) and N-acetyl-α-1,4-polygalactosamine (NAPGA) (Chapter 16). These low toxic and biodegradable polymers are capable of stimulating the host immune system and inhibiting the growth of certain tumor cells. The affinity for hepatocytes is another advantage of these polysaccharides which makes them suitable for site-directed delivery. A water soluble macromolecular prodrug of 5-fluorouracil with affinity for tumor cells and without the usual side-effects was obtained which demonstrated an enhanced activity, compared to the free 5FU, against HLE hepatoma cells *in vitro*.

Simulation of Polymer-Solvent Interactions

The process of drug delivery and controlled drug release takes place in an aqueous environment. As has been shown previously, hydrogels play an important role in polymeric drug delivery systems. An effective controlled release system is supposed to deliver a desired amount of drug to the treatment site at a certain rate. The rate of drug delivery depends on the interactions between the polymeric material and the solvent, and on the ability of the polymer to swell. The aqueous swelling is dependent on the hydrophilicity of the polymer, the structure of the hydrogel network formed, and the amount of ionizable groups in the system *(82)*. The swelling of hydrogels is important both for drug loading into the system and for drug release. The understanding of interactions occurring between polymers and water and subsequent hydrogel swelling is important for the development of new advanced drug delivery systems. These mechanisms can be studied both experimentally and theoretically.

Computer simulations could be a useful method to study the properties and behavior of systems at molecular level. Recently, the computer simulations have been applied to polymer systems *(83)*. The molecular mechanics studies are based on a mathematical model that represents a potential energy surface for the molecule of interest. The potential energy surface describes the energy of the system as a function of the three-dimensional structure of the molecules. The energetics, structural, and dynamic aspects of a system can be investigated. The computational techniques allow for simulating systems which include 10,000 to 50,000 atoms with realistic mathematical models.

In Chapter 18, a system of poly(ethylene oxide) hydrogel in water is chosen as a model system. The use of different dielectric constants to simulate the effect of the solvent results in two distinctly different PEO conformations in water and in nonpolar solvents. The PEO polymer chain has a helical structure in water while it has a random coil structure in nonpolar solvents *(84,85)*. However, the unique properties of water, such as the small size, highly variable orientations, and the ability to form directional interactions, make the simulation based only on dielectric constants inadequate. To understand the mechanism of polymer-solvent interactions at the molecular level, the explicit addition of water molecules is required. The simulation with explicit inclusion of water molecules into the system has demonstrated the effect of water on the conformation of the hydrophilic and hydrophobic polymer chains and resulted in a more accurate description of polymer-water interaction at molecular level.

Literature Cited

1. "Medical Uses of Plastics to Grow 6% Annually", *C&E News* 7, 10, 1989.
2. "High Performance Biomaterials", Szycher, M., ed., Technomic, Lancaster, 1991, p.47..
3. Urry, D. W. *Prog. Biophys. Molec. Biol.* **1992**, 57, 23-57.
4. Urry, D. W. *Angew. Chem. Int. Ed. Engl.* **1993**, June issue.
5. Urry, D. W. In *Cosmetic and Pharmaceutical Applications of Polymers;* Gebelein, C. G; Cheng, T. C.; Yang, V. C., Eds.; Plenum Press, New York, 1991; 181-192.
6. Urry, D. W.; Gowda, D. C.; Parker, T. M.; Luan, C-H.; Reid, M. C.; Harris, C. M.; Pattanaik, A.; Harris, R. D., *Biopolymers* **1992**, 32, 1243-1250.
7. (a) Krowicki, K.; Lee, M.; Hartley, J.A.; Ward, B.; Kissinger, K.; Skorobogaty, A.; Dabrowiak, J.C.; Lown, J.W. *Structure & Expression* **1988**, 2, 251. (b) Zimmer, C.; Wahnert, U. *Prog. Biophys. Molec. Biol.* **1986**, 47, 31. (c) Dervan, P.B. *Science* **1986**, 234, 464.

8. Lown, J.W. *Anti-Cancer Drug Des.* **1988**, 3, 25.
9. Karlin, S. *Proc. Natl. Acad. Sci. USA* **1986**, 83, 6915.
9. Hahn, F.E. In *Antibiotics III. Mechanisms of Action of Antimicrobial and Antitumor Agents.* Corcoran, J.W and Hahn, F.E. Eds. Springer-Verlag, N.Y., **1975**, 79.
10. Kopka, M.L.; Yoon, C.; Goodsell, D.; Pjura P.; Dickerson, R.E. *Proc. Natl. Acad. Sci. USA* **1985**, 82, 1376.
11. Lown, J.W.; *Biochem.* **1986**, 25, 7408.
12. Kissinger, K.; Krowicki, K.; Dabrowiak, J.C.; Lown, J.W. *Biochem.* **1987**, 26, 5590.
13. Lee, M.; Hartley, J.A.; Pon, R.T.; Krowicki K.; Lown, J.W. *Nucleic Acids Res.* **1987**, 16, 665 and references therein.
14. Lown, J.W. In *Molec. Basis of Specificity in Nucleic Acid-Drug Interactions.* Pullman, B., Jortner, J., eds. Kluwer Acad. Pub., Netherlands, 1990, p. 103 and references therein.
15. Ito, Y. In *Synthesis of Biocomposite Material: Chemical and Biological Modifications of Natural Polymers.* Imanishi, Y., Ed., CRC Press, Boca Raton, 1992, 137-180.
16. Ito, Y., Casolaro, M., Kono., K., Imanishi, Y. *J. Controlled Release* **1989**, 10, 195-203.
17. Ito, Y., Kotera, S., Inaba, M., Kono., K., Imanishi, Y. *Polymer* **1990**, 31, 2157-2161.
18. Ito, Y., Inaba, M., Chung, D.J., Imanishi, Y. *Macromolecules* **1992**, 25, 7313-7316.
19. Ito, Y., Chung, D.J., Imanishi, Y. *Artif. Organs* **1990**, 14, 234-236.
20. Chung, D.J., Ito, Y., Imanishi, Y. *J. Controlled Release* **1992**, 18, 45-54.
21. Heller, J.; Pangburn, S. H.; Roskos, K. V. *Biomaterials* **1990**, 11, 345-350.
22 "Biodegradable Hydrogels for Drug Delivery", Park. K., Shalaby, W.S.W., and Park, H., eds. Technomic, Lancaster-Basel, 1993, p.106-108.
23. Artursson, P.; Edman, P.; Laakso, T.; Sjöholm, I. *J. Pharm. Sci.* **1984**, 73, 1507-1513.
24. Edman, P.; Ekman, B.; Sjöholm, I. *J. Pharm. Sci.* **1980**, 69, 838-842.
25. Park, K. *Biomaterials* **1988**, 9, 435-441.
26. Guiseley, K. B. In *Industrial Polysaccharides: Genetic Engineering, Structure/property Relations and Applications,* Yalpani, M., Ed.; Elsevier Science Publishers B. V.: Amsterdam, 1987; pp 139-147.
27. Plate, N. A.; Malykh, A. V.; Uzhinova, A. D.; Panov, V. P.; Rozenfel'd, M. A. *Polymer Sci. USSR* **1989**, 31, 220-226.
28. Seller, K., Szathmary, S., Huss, J., DeCoster, R., Junge, W. In *Minutes of the Fifth International Symposium on Cyclodextrins,* Dominiques Duchene, Ed., Editions De Sante, 1990, 518-521.
29. Coussement, W., Van Cauteren, H., Vandenberge, J., Vanparys, P., Teuns, G., Lampo, A., Marsboom, R. In *Minutes of the Fifth International Symposium on Cyclodextrins,* Dominiques Duchene, Ed., Editions De Sante, 1990, 522-524.
30. Brewster, M., Bodor, N. In *Minutes of the Fifth International Symposium on Cyclodextrins,* Dominiques Duchene, Ed., Editions De Sante, 1990, 525-534.
31. Hepel, M. *Proceedings of the Third International Meeting on Chemical Sensors,* Cleveland, Ohio, September 24-26, 1990, pp. 35-37.

32. Zhou, Q.X., Miller, L.L., and Valentine, J.R. *J. Electroanal. Chem.* **1989**, 261, 147-164.
33. Miller, L.L., and Zhou, Z. X. *Macromolecules* **1987**, 20, 1594-1597.
34. "Concise Encyclopedia of Polymer Science and Engineering", Kroschwitz, J.I., Wiley & Son, New York, 1990, p.673-674.
35. Costall, B., Domeney, A.M., Naylor, R.J. *Br. J. Pharmacol.* **1981**, 74, 899P-900P (Abstr.)
36. Bodor, N., Brewster, M.E. *Pharmacol. Ther.* **1983**, 19, 337-386.
37. Siew, C., Goldstein, D.B. *J. Pharmacol. Exp. Ther.* **1978**, 204, 541-46.
38. Kroin, Y.S., Penn, R.D. *Neurosurgery* **1982**, 10, 349-354.
39. de la Torre, J. C., Gonzalez-Carvajal, M. *Lab. Anion. Sci.* **1981**, 31, 701-703..
40. Kwon, I.C., Bae, Y.H., Okano, T., and Kim, S.W. *J. Control. Rel.* **1991**, 17, 149-156.
41. Kwon, I.C., Bae, Y.H., and Kim, S.W. *Nature* **1991**, 354, 291-293.
42. Bae, Y.H., Kwon, I.C., Pai, C.M., and Kim, S.W. *Makromol. Chem., Macromol. Symp.* (in press).
43. Hopfenberg, H. B.; Hsu, K. C. *Polym. Eng. and Sci.* **1978**, 18, 1186.
44. Hopfenberg, H. B.; Apicella, A.; Saleeby, D. E. *J. of Membrane Sci.* **1981**, 8, 273.
45. Peppas, N. A.; Franson, M. N. *J. Polym. Sci., Polym. Phys. Ed.* **1983**, 21, 983.
46. Korsmeyer, R. W.; Peppas, N. A. *J. Controlled Release* **1984**, 1, 89.
47. Gaeta, S.; Apicella, A.; Hopfenberg, H. B. *J. Membrane Sci.* **1982**, 12, 195.
48. Good, W. R.; Mueller, K. F. *AIChE Symp. Ser.* **1981**, 77, 42.
49. Lee, P. I. *Polymer Comm.* **1983**, 24, 45.
50. Sefton, M. V.; Brown, L. R.; Langer, R. S. *J. Pharm. Sci.* **1984**, 73 (12), 1859.
51. Hunt, G.; Kearney, P.; Kellaway, I.W. In *Drug Delivery Systems*; P. Johnson and J. G. Lloyd-Jones, Eds.: Ellis Horwood Ldt. Chichester and VHC Verlagsgesellschaft GmbH Weinheim, 1987, 180-199.
52. Sau-Hung; Leungand, S.; Robinson, J.R. *Polymer News* **1990**, 15, 333.
53. Chen, J.L.; Cyr, G.N. In *Adhesive Biological Systems*, R.S. Manly, Ed. Academic Press, New York and London, 1970.
54. Wilhelm, M., Zhao, C.-L., Wang, Y., Xu, R., Winnik, M.A., Mura, J.-L., Riess, G., Croucher, M.D. *Macromol.* **1991**, 24, 1033.
55. Yokoyama, M., Miyauchi, M., Yamada, N., Okano, T., Sakurai, Y., Kataoka, K., Inoue, S. *Cancer Res.* **1990**, 50, 1693-1700.
56. Butler, G. B. In *Anionic Polymeric Drugs: Synthesis, Characterization and Biological Activity of Pyran Copolymers*, John Wiley & Sons; p.49, 1980.
57. Ottenbrite, R. M. In *Anionic Polymer Drugs: Structure and Biological Activities of Some Anionic Polymers*; John Wiley & Sons: New York; p.21, 1980.
58. Ottenbrite, R. M., Kuus, K., and Kaplan, A. M. *Polymer Preprints* **1983**, 24(1), 25.
59. Ottenbrite, R. M., Kuus, K., Kaplan, A. M. In *Polymer in Medicine*; Plenum Press: New York, pp.3-22, 1984.
60. Ottenbrite, R.M., Kuus, K., Kaplan, A.M. *J. Macromol. Sci.-Chem.* **1988**, A25, 873-893.

61. Ottenbrite, R. M., Takatsuka, R. *J. of Bioactive and Compatible Polymers* **1986**, 1, 46.
62. Claes, P., Billiau, A., DeClercq, E., Desmyter, J., Schonne, E., Vanderhaeghe, H., and Desomer, P. *J. Virol.* **1970**, 5, 313.
63. Schuller, B. B., Morahan, P.S., and Snodgrass, M. *Cancer Res.* **1975**, 35, 1915.
64. Duncan, R., Kopecek, J. In *Advances in Polymer Science: Soluble Synthetic Polymers as Potential Drug Carries*, New York and Berlin, 57, p.51, 1984.
65. Sato, T., Kojima, K., Ihda, T., Sunamoto, J., and Ottenbrite, R.M. *J. Bioact. Compat. Polym.* **1986**, 1, 448.
66. Suda, Y., Yamamoto, H., Sumi, M., Oku, N., Ito, F., Yamashita, S., Nadai, T., and Ottenbrite, R.M. *J. Bioact. Compat. Polym.* **1991**, 7.
67. Boddy, A.; Aarons, L. *Advanced Drug Delivery Reviews* **1989**, 3(2), 155-266.
68. Goddard, P. *Advanced Drug Delivery Reviews* **1991**, 6(2), 103-233.
69. Kamata, R.; Yamamoto, T.; Matsumoto, K.; Maeda, H. *Infect. Immunity* **1985**, 48, 747-753.
70. Vert, M.; *Angew. Makromol. Chem.* **1989**, 166-167, 155.
71. Gilding, D.K., In *Biocompatibility of Clinical Implant Materials*, vol. II; Williams, D.F. Ed.; CRC Press Inc., Boca Raton, Florida, 1981, p 218.
72. Kulkarni, R.K.; Pani, K.C.; Neuman, C.; Leonard, F. *Arch. Surg.* **1966**, 93, 839.
73. Doddi. N.; Versfeldt, C.C.; Wasserman, D., US patent 4,052,988, 1977.
74. Casey, D.J.; Roby, M.S., Eur. Patent EP 098 394 A1,1984.
75. Asano, M.; Yoshida, M.; Kaetsu, Katakai, R.; Imai, K.; Mashimo, T.; Yuasa, H.; Yamanaka, H. *J. Jpn. Soc. Biomat.* **1985**, 3, 85.
76. Asano, M.; Yoshida, M.; Kaetsu, Katakai, R.; Imai, K.; Mashimo, T.; Yuasa, H.; Yamanaka, H. *Seitai Zairyo* **1986**, 4, 65.
77. Kaetsu, I.; Yoshida, M.; Asano, M.; Yamanaka, H.; Imai, K.; Yuasa, H.; Mashimo, T.; Suzuki, K.; Katakai, R.; Oya, M. *J. Controlled Rel.* **1986**, 6, 249.
78. Schakenraad, J.M.; Nieuwenhuis, P.; Molenaar, I.; Helder, J.; Dijkstra, P.J.; Feijen, J. *J. Biomed. Mater. Res.* **1989**, 23, 1271.
79. Helder, J.; Dijkstra, P.J.; Feijen, J. *J. Biomed. Mater. Res.* **1990**, 24, 1005.
80. Barrows, T.H.; Johnson, J.D.; Gibson, S.J.; Grussing, D.M. In *Polymers in Medicine II*, Chiellini, E., Giusti, P.,Migliaresi, C., Nicolais, E., Eds.; Plenum Press: New York, 1986; p.85.
81. Horton, V.L.; Blegen, P.E.; Barrows, T.H.; Quarfoth, G.L.; Gibson, S.J.; Johnson, J.D.; McQuinn, R.L. In *Progress in Biomedical Polymers*, Gebelein, C.G., Dunn, R.L., Eds.; Plenum Press: New York, 1990; p.263.
82. Gehrke, S. H.; Lee, P. I. In *Specialized Drug Delivery Systems: Manufacturing and Production Technology*; Tyle, P., Ed.; Marcel Dekker, Inc.: New York, 1990; Ch. 8.
83. Roe, R. J., Ed.; *Computer Simulation of Polymers*; Prentice Hall: Englewood Cliffs, 1991.
84. Bailey, F. E., Jr.; Lundberg, R. D.; Callard, R. W. *J. Polym. Sci. Part A* **1964**, 2, 845-851.
85. Maron, S. H.; Filisko, F. E. *J. Macromol. Sci. Phys.* **1972**, 6, 79-90.

RECEIVED August 23, 1993

Chapter 2

Bioelastic Materials and the ΔT_t-Mechanism in Drug Delivery

Dan W. Urry[1], D. Channe Gowda[1,2], Cynthia M. Harris[1,2], and R. Dean Harris[1,2]

[1]Laboratory of Molecular Biophysics, University of Alabama at Birmingham, VH300, Birmingham, AL 35294–0019
[2]Bioelastics Research, Ltd., 1075 South 13th Street, Birmingham, AL 35205

Bioelastic matrices are capable of effecting drug release by means of diffusion, degradation, transduction or a designed combination of any of the three. With proper design the transductional release can utilize free energy inputs including changes in the intensive variables of mechanical force, temperature, pressure, chemical potential, electrochemical potential and electromagnetic radiation. In the present report, the feasibility of chemical clocks to introduce both degradational and transductional release is demonstrated, and diffusional release from a limiting case of a thin disc of the simplest biocompatible matrix is described using the drug biebrich scarlet which has wound repair efficacy. A local release of 50 to 15 nanomoles/day of biebrich scarlet could be sustained for almost a week. This biocompatible matrix, X^{20}-poly(GVGVP), has been found not to become coated with a fibrous capsule even after months of implantation within the peritoneal cavity and under the conjunctiva of the eye. It is suggested that larger constructs utilizing chemical clocks could result in sustained release of the order of 1 μmole per day for much of a year.

Living organisms develop and survive by achieving the capacity to convert available energy sources into motion and other functions required for life. This entails free energy transductions involving the intensive variables of mechanical force, temperature, pressure, chemical potential, electrochemical potential and electromagnetic radiation (*e.g.*, visible and ultraviolet light). A common molecular mechanism whereby these energy conversions can occur is the ΔT_t-mechanism which utilizes inverse temperature transitions leading to increased order by hydrophobic folding and assembly of protein on raising the temperature in water and by changing the temperature, T_t, at which the inverse temperature transition occurs using one of the above intensive variables to lower the value of T_t from

0097–6156/94/0545–0015$08.00/0

above to below physiological temperature and thereby to drive hydrophobic folding and assembly.

The ΔT_t-Mechanism *(1,2)*

The ΔT_t-mechanism was discovered and developed using elastomeric polypeptides, a class of protein-based polymers called bioelastic materials; it is considered relevant to protein folding and function. The demonstrated and putative energy conversions, to date, are given in Figure 1. As represented in Figure 2a, poly(GVGVP) is extended at low temperature and hydrophobically folds on raising the temperature with the capacity to lift a weight and perform work. This is thermomechanical transduction. Now if the transition occurs as in curve **b** and an

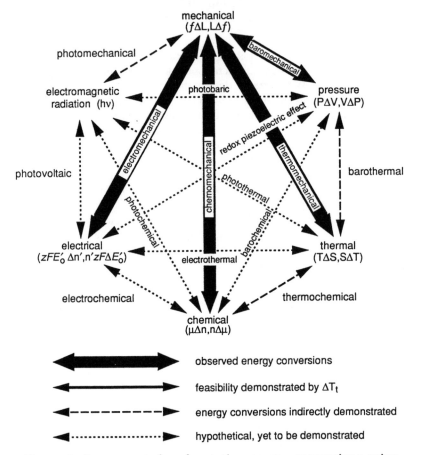

Figure 1. Demonstrated and putative energy conversions using bioelastic materials and the ΔT_t - mechanism.

intensive variable is introduced which lowers the transition as in curve **a**, then complete contraction is achieved when the temperature is within the range indicated by **c**. This is the ΔT_t-mechanism. Methods whereby the value of T_t can be changed are shown in Figure 2b (ii-vi). Shown in Figure 2c is the elastic model protein bands achieved by 20 Mrad gamma-irradiation cross-linking of poly(GVGVP), or a suitable modification, undergoing contraction (a) or relaxation (b) by the means indicated in Figure 2d. Contraction/relaxation can be a deswelling/swelling process with a volume change in water of a factor of 10.

Biocompatibility of Bioelastic Materials *(3)*

Employing eleven recommended biological tests for determining biocompatibility, the basic materials [poly(Gly-Val-Gly-Val-Pro) and its 20 Mrad gamma-irradiation cross-linked matrix, X^{20}-poly(GVGVP)] have been demonstrated to exhibit remarkable biocompatibility *(3)*. Accordingly, these materials have become candidates for effecting controlled release of pharmaceuticals. In this regard, it is significant, when implanted in the rat and the rabbit for periods of up to several months, that there is no fibrous capsule formation around X^{20}-poly(GVGVP) *(4,5)*. This allows that the matrix itself can control the rate of release without the limitations imposed by an encapsulating fibrous protein coat. After months of being implanted within the peritoneal cavity or under the conjunctiva of the eye, X^{20}-poly(GVGVP) remains a soft, transparent, elastic matrix which retains its capacity to exhibit transductional behavior.

The contracted transparent elastic matrix, X^{20}-Poly(GVGVP), in spite of being formed by means of hydrophobic folding and assembly, is 50% water by weight under physiological conditions *(6-8)*. It can be loaded with agents simply by swelling in the presence of the agent dissolved in water and then by contracting to expel excess water. X^{20}-poly(GVGVP), for example, can increase some 10-fold in volume on lowering the temperature from 37°C to 20°C. This swell-doping is particularly effective for agents with relatively hydrophobic moieties and limited solubility in water such as biebrich scarlet and methylene blue. These properties also suggest useful loading with pharmaceuticals such as steroids, peptides (*e.g.*, enkephalin), oligonucleotides and even small proteins.

Controlled Release

Diffusional release is a mechanism available to essentially all drug-laden matrices. Degradational release of agents is possible with a more limited subset of matrices including those comprised of polypeptides and polyesters. The bioelastic polymers can also be synthesized with periodic esters in the backbone but also can be designed to contain sites for proteolytic enzymes anticipated in the milieu of the implant site or even for proteolytic enzymes doped within the matrix itself.

Perhaps most significantly, the present materials introduce a versatile set of transductional release processes that have the potential for increased controlled release. For example, there can be introduced chemical clocks such as the

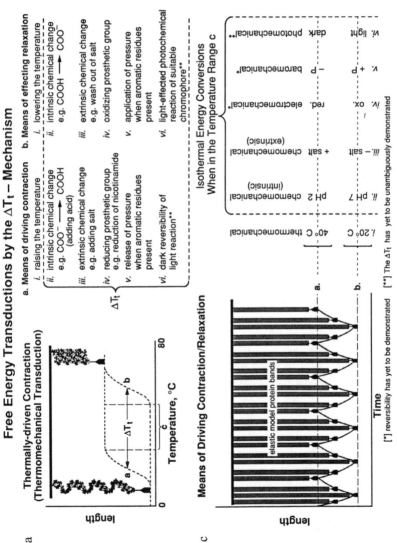

Figure 2. Summary of contractions and relaxations (swelling) achievable with bioelastic materials designed for input energies of thermal, chemical, pressure, electrochemical and light.

carboxamides of asparagine or glutamine residues wherein, depending on the sequence in which the residue occurs, the carboxamide will hydrolytically cleave at the matrix-milieu interface with a particular half-life *(9)*. With a properly designed matrix this can result in raising T_t above physiological temperature with the consequences of dramatic swelling, enhanced diffusional release and enhanced degradation *(10)*. Introduction of an occasional ester in the backbone by replacing a glycine residue with a glycolic acid residue allows for chain breakage at the matrix interface, and also results in the formation of carboxylate which would enhance swelling. Changes in any of the intensive variables of temperature, pressure, chemical potential, electrochemical potential and electromagnetic radiation could be used to control the rate of release by swelling and degradation or conversely to control the rate of release by driving contraction of an envelope to expel contents *(1,2)*.

Model Elastic Protein-based Polymers for Controlled Release

As an initial effort at introducing transductional capacities into controlled release, a series of sequential polypeptides have been prepared. These are: **I**, poly(GVGVP); **II**, poly[0.85(GVGVP), 0.15(GVG_cVP)]; **III**, poly[0.90(VPGVG), 0.10(VPG_cVG)]; and **IV**, poly[0.94(VPGVG), 0.06(VPG_cNG)] where V = Val, P = Pro, G = Gly, G_c = glycolic acid, and N = Asn (asparagine). The rates of breakdown were followed using the time dependence of the temperature, T_t, of the inverse temperature transition when in 0.2 N phosphate to maintain pH in the 7.0 to 7.5 range. Using the 8-day time interval from 6 to 14 days which is quite linear for all polymers, the $\Delta T_t/8$ days was 0 for **I**, 2.7 °C for **II**, 1.6 °C for **III**, and 3.2 °C for **IV**. As these polymers contained different mole fractions for the pentamer having the chemical clock, the values were normalized for that variable. This gave 0°, 18°, 16°, and 53 °C, respectively. Accordingly, there appeared to be little difference in the rate of breakdown of the glycolic ester whether in VG_cV or the PG_cV sequence. However, the combination of a carboxamide in the side chain and an ester in the backbone increased the rate for swelling and breakdown some three-fold.

Further, beginning efforts at controlled release have involved swell-doping of X[20]-poly(GVGVP) matrices in the form of thin discs with biebrich scarlet and methylene blue. Both of these agents are taken very well into the matrix, and, after initial rinsing and an initial burst, there is a period approaching a week where release is sustained in the 50 nanomole to 15 nanomole/day range.

Experimental

Polypentapeptide Syntheses. The syntheses of VPGVG and GVGVP have been previously described *(11,12)*. For conformational reasons, that is, due to the Pro[2]-Gly[3] Type II ß-turn, the usual numbering of the pentamer is Val[1]-Pro[2]-Gly[3]-Val[4]-Gly[5] *(1,12)*. The pentamers containing glycolic acid (G_c) and asparagine (N) were synthesized as VPG_cVG, VPG_cNG and GVG_cVP by the solution phase method. It has been found in our previous studies that the highest molecular weights were obtained when the pentamer permutation with Pro at the carboxyl terminus (GVGVP) and para-nitrophenol (ONp) were used for carboxyl activation *(12)*. Here, however, the VPGVG permutation was used with Gly at the C terminus for pentamers substituted in position 3 with the glycolic acid residue and with 1-ethyl-3-dimethylaminopropylcarbodiimide (EDCI) as the polymerizing agent. The reasons for these changes were (1) with GVGVP permutation, glycolic acid at position 3 occurs at the N terminus which may be difficult to polymerize under normal conditions necessitating in these cases the use of the VPGVG permutations, and (2) at the end of the polymerization, the terminus ONp moieties were removed by base treatment with 1 N NaOH which would hydrolyze ester bonds present in the polymers. Moreover, our previous studies *(3,13)* indicate that using EDCI with Pro at the carboxyl terminus also gave equally good polymers which have identical transition temperatures, carbon-13 nuclear magnetic resonance (NMR) spectra, amino acid analyses and biological studies. When the VPGVG permutation was used with ONp, the temperature, T_t, of the inverse temperature transition was higher.

The peptides Boc-Val-Pro-OBzl, Boc-Val-Gly-OBzl, Boc-Val-Pro-Gly-Val-Gly-OH and Boc-Gly-Val-Gly-Val-Pro-OH were prepared as previously described **(11,12)**. Pentamer purity prior to polymerization is a critical factor in obtaining high molecular weight polymers in good yield as impurities can result in termination of the polymerization process. The protected peptides were characterized by carbon-13 nuclear magnetic resonance before polymerization to verify the structure and purity. Thin layer chromatography (TLC) was performed on silica gel plates obtained from Whatman, Inc., with the following solvent systems: R_f^1, $CHCl_3$:CH_3OH:CH_3COOH (95:5:3); R_f^2, $CHCl_3$:CH_3OH:CH_3COOH (90:10:3); R_f^3, $CHCl_3$:CH_3OH:CH_3COOH (85:15:3). The compounds on TLC plates were detected by UV light by spraying with ninhydrin, or by chlorine/tolidine spray. All Boc amino acids, N,N-diisopropylcarbodiimide and HOBt were purchased from Advanced Chem. Tech. (Louisville, Kentucky). EDCI was obtained from Bachem, Inc. (Torrance, California). Glycolic acid and carbonyldiimidazole were purchased from Aldrich Chemical Company (Milwaukee, WI).

N,N-Diisopropyl-O-Benzylisourea. Benzyl alcohol (42.7 g, 0.4 mol) was added with stirring to a mixture of cuprous chloride (0.15 g) in N,N-diisopropylcarbodiimide (50.0 g, 0.4 mol) over a period of 30 min at $0°C$. After an additional 1 h at $0°C$, the reaction was stirred at room temperature for 18 h to ensure complete reaction. The volume was then doubled with hexane and the solution was applied to a filter pad of neutral alumina to remove copper salts. The product was eluted with a total volume of 1 L of hexane, and the solvent was evaporated under reduced pressure and dried.

Benzyl-Glycolic acid *(1)*. Glycolic acid (7.61 g, 0.1 mol) was added to N,N-diisopropylbenzylisourea (23.3 g, 0.1 mol) with stirring. The mixture became very viscous within 10 min and was stirred intermittently for 1 h. The volume was then increased to 200 mL with THF, and the mixture was stirred for 48 h at room temperature. After cooling the mixture to -15 °C, the diisopropylurea was removed by filtration, and the THF was evaporated under reduced pressure to give 15.26 g (yield, 91.8%) of benzyl glycolate.

Boc-Val-Pro-Glc-OBzl (2). A solution of Boc-Val-Pro-OH (9.43 g, 0.03 mol) in dry methylene chloride (50 mL) was cooled to 0 °C with stirring and treated with a solution of carbonyldiimidazole (4.86 g, 0.03 mol) in methylene chloride (50 mL) over 30 min. After stirring for an additional 20 min at 0 °C, benzyl glycolate (5.0 g, 0.03 mol) was added over a 30 min period. The reaction mixture was maintained at 0 °C for a further 2 h, and then at room temperature for 3 days. The mixture was evaporated to a thick oil which was dissolved in chloroform and extracted with water, 20% citric acid, water, saturated sodium bicarbonate, water, and dried over sodium sulfate. The solvent was removed under reduced pressure and the resulting oil was recrystallized from ether/petroleum ether to obtain 9.0 g (yield, 64.84%) of 2: R^1_f, 0.48; R^2_f, 0.59.

Boc-Asn-Gly-OBzl (3). A mixture of Boc-Asn-ONp (3.53 g, 0.01 mol), Gly-OBzl.p-Tosylate (3.77 g, 0.01 mol) and HOBt (1.35 g, 0.01 mol) in DMF (50 mL) was stirred at room temperature for 2 days maintaining the pH at 7.5 to 8.0 with NMM. The solvent was removed under reduced pressure and worked up by acid-base extraction. The resulting gum was recrystallized from ether to obtain 2.8 g (yield, 73.68%) of 3: R^1_f, 0.26; R^2_f, 0.35.

Boc-Val-Pro-Glc-Val-Gly-OBzl (4). Compound 2 (9.0 g, 0.0195 mol) was dissolved in dry THF (100 mL), and 10% palladium on activated charcoal (1.0 g) was added. This mixture was hydrogenated at 50 psi for 6 h and the catalyst was removed by filtration through celite. The filtrate was evaporated *in vacuo*, and dried to give the acid, Boc-Val-Pro-Glc-OH.

Boc-Val-Gly-OBzl (3.3 g, 0.009 mol) was deblocked by stirring for 1.5 h in 4 N HCl in dioxane. Excess HCl and dioxane were removed under reduced pressure, triturated with ether, filtered, washed with ether, and dried.

A solution of Boc-Val-Pro-Glc-OH (3.35 g, 0.009 mol) and HOBt (1.35 g, 0.01 mol) in DMF was cooled to -15 °C with stirring, and EDCI (1.97 g, 0.01 mol) was added. After 20 min, a pre-cooled solution of the above hydrochloride salt and NMM (0.99 mL, 0.01 mol) was added and the reaction mixture was stirred overnight at room temperature. The mixture was evaporated to a thick oil which was dissolved in chloroform. This solution was extracted with water, 10% citric acid, water, 5% sodium bicarbonate, water, and dried over sodium sulfate. The solvent was removed under reduced pressure, and the resulting oil was dissolved in ether and precipitated from petroleum ether. The solid was filtered, washed with petroleum ether, and dried to obtain 4.1 g (yield, 73.61%) of 4: R^2_f, 0.51; R^3_f, 0.64.

Boc-Val-Pro-Glc-Asn-Gly-OBzl (5). Compound 3 (2.9 g, 0.0075 mol) was deblocked with trifluoroacetic acid and coupled to Boc-Val-Pro-Glc-OH (2.75 g, 0.0074 mol) using EDCI with HOBt in the same manner as that described for **4** to give 2.82 g (yield, 60.1%) of **5**: R^2_f 0.33; R^3_f 0.41.

Boc-Val-Glc-OBzl (6). Compound **1** (7.5 g, 0.04 mol) was coupled to Boc-Val-OH (9.6 g, 0.04 mol) using carbonyldiimidazole (7.3 g, 0.045 mol) following the same procedure described for the preparation of **2** to give 12.5 g (yield, 76.05%) of **6**: R^1_f 0.55; R^2_f 0.65.

Boc-Gly-Val-Glc-OBzl (7). Compound **6** (8.04 g, 0.022 mol) was deblocked using HCl/dioxane and coupled to Boc-Gly-OH (3.5 g, 0.02 mol) using EDCI with HOBt as described in the preparation of **4** to obtain 7.5 g (yield, 88.7%) of **7**: R^1_f, 0.44; R^2_f, 0.52.

Boc-Gly-Val-Glc-Val-Pro-OBzl (8). Compound **7** (4.65 g, 0.011 mol) was hydrogenated into free acid, then coupled to HClVP-OBzl (obtained by deblocking 4.45 g of Boc-Val-Pro-OBzl with HCl/dioxane) using EDCI in the presence of HOBt and following the same procedure as described for **4** to obtain 6.0 g (yield, 91.4%) of **8**: R^1_f 0.21; R^2_f 0.29.

Poly(Gly-Val-Gly-Val-Pro) (I). Boc-Gly-Val-Gly-Val-Pro-OH (2.11 g, 0.004 mol) was deblocked with TFA, and a one-molar solution of TFA salt in DMSO was polymerized for 12 days using EDCI (2 equivalents) as the polymerizing agent with HOBt (1 equivalent) and 1.6 equivalents of NMM as base. The polymer was dissolved in water, dialyzed using 3500 Mw cut-off dialysis tubing for one week and lyophilized. It was then dialyzed using 50 kD Mw cut-off dialysis tubing for one week and lyophilized to obtain 1.09 g (yield, 66.84%) of **I**.

Poly[0.9(Gly-Val-Gly-Val-Pro), 0.1(Gly-Val-Glc-Val-Pro)] (II). Compound **8** (6.0 g, 0.0096 mol) was dissolved in THF (60 mL), and 10% palladium on charcoal (0.6 g) was added. This mixture was hydrogenated at 40 psi for 6 h; the resulting residue was triturated with ether, filtered, and dried to obtain the acid.
Boc-Gly-Val-Gly-Val-Pro-OH (10.55 g, 0.02 mol) and Boc-Gly-Val-Glc-Val-Pro-OH (1.18 g, 0.0022 mol) were deblocked together with TFA and polymerized using the same procedure as for compound **I** to obtain 4.3 g (yield, 47.3%) of **II**.

Poly[0.9(Val-Pro-Gly-Val-Gly), 0.1(Val-Pro-Glc-Val-Gly)] (III). Compound **4** (4.1 g, 0.0066 mol) was hydrogenated as described above to obtain 3.1 g (yield, 88.57%) of the acid. This acid (1.32 g, 0.0025 mol) was mixed with Boc-Val-Pro-Gly-Val-Gly-OH (11.87 g, 0.022 mol) and deblocked with TFA. The polymerization, dialysis and lyophilization were carried out using the same procedure described for polymer **I** to obtain 4.1 g (yield, 40.04%) of **II**.

Poly[0.9(Val-Pro-Gly-Val-Gly), 0.1(Val-Pro-Glc-Asn-Gly)] (IV). Compound **5** (2.8 g, 0.0044 mol) was debenzylated following the same procedure described above to obtain 1.8 g (yield, 75%) of the acid. This acid (1.35 g, 0.0025 mol) and Boc-Val-Pro-Gly-Val-Gly-OH (11.87 g, 0.022 mol) were deblocked, polymerized, dialyzed and lyophilized using the same procedure as described for polymer I to obtain 3.8 g (yield, 37.3%) of IV.

The purity and the composition of the final products was checked by carbon-13 nuclear magnetic resonance and amino acid analyses. Based on the amino acid analyses, the more correct statement of the formulae for the polymers with mixed pentamers would be II: poly[0.85(GVGVP), 0.15(GVG$_c$VP)], III: poly[0.9(VPGVG), 0.1(VPG$_c$VG)], and IV: poly[0.94(VPGVG), 0.06(VPG$_c$NG)].

Abbreviations: Boc, tert-butyloxycarbonyl; ONp, para-nitrophenol; OBzl, benzyl; EDCI, 1-ethyl-3-dimethylaminopropylcarbodiimide; HOBt, 1-hydroxybenzotriazole; TFA, trifluoroacetic acid; NMM, N-methylmorpholine; DMSO, dimethylsulfoxide; DMF, dimethylformamide; THF, tetrahydrofuran; V (Val), valine; P (Pro), proline; G (Gly), glycine; N (Asn), asparagine; G$_c$ (Glc), glycolic acid.

Determination of T$_t$. Below a certain temperature each of the polymers I though IV is soluble in water. On raising the temperature, there is the onset of aggregation resulting in a phase transition to form a more-dense, polymer-rich, viscoelastic state called a coacervate. The onset of the transition can be followed spectrophotometrically by the temperature profile of turbidity formation where the temperature for half-maximal turbidity is designated as T$_t$, the temperature of this inverse temperature transition in which increased hydrobicity lowers the value of T$_t$ and increased polarity as the formation of carboxylates markedly increases T$_t$.

For these elastic protein-based polymers, it has been found that 40 mg/mL is the high concentration limit above which increasing concentration no further lowers the value of T$_t$. Accordingly, all polymers were studied beginning with concentrations of 40 mg/mL. The polymers were dissolved in 0.2 N phosphate at pH 7.5 in order to maintain a near constant pH during the breakdown of carboxamides and esters to polar carboxylates. The value of T$_t$ for each polymer was determined at zero time and then the samples were incubated at 37°C in a rocker device for 24 h and the temperature lowered to achieve dissolution and a new determination of T$_t$. This process was repeated each 24 h for 21 days. The increases in T$_t$ with time at 37°C were due to hydrolysis of carboxamides and esters. The temperature profiles of turbidity formation were run on a Pye-Unicam 8610 spectrophotometer at 400 nm.

Formation of the X^{20}-poly(GVGVP) Bioelastic Matrix. The coacervate state of poly(GVGVP) was placed in a mold capable of forming sheets 0.35 mm in thickness and cross-linked at the Auburn University Nuclear Science Center with a 20 Mrad dose of gamma-irradiation. Discs of diameter 5.45 mm and a thickness of 0.35 mm were punched from the bioelastic sheet equilibrated in water at 37°C.

Loading the Discs with Biebrich Scarlet and Methylene Blue. A 47 μL, 0.1 M solution of biebrich scarlet and of methylene blue were each added to a disc of volume 8.2 μL. As the 37°C state of X^{20}-poly(GVGVP) can expand in water on lowering the temperature to take up a 10-fold volume of water, a 5 to 6-fold volume was entirely taken into the disc and the swollen disc was allowed to stand overnight at 4°C. The temperature was then raised to 37°C causing the extrusion of water and excess biebrich scarlet. Of the 4.7 μmoles of biebrich scarlet in the first experiment, 2.5 μmoles were retained within the 1 volume of contracted disc, and 2.2 μmoles were released to the 5.7 volumes of extruded water. In the second experiment, the values were 1.7 and 3.0 μmoles, respectively. Thus, the swell-doping occurred with a several-fold greater partitioning into the disc of X^{20}-poly(GVGVP). In the first experiment with a longer time at 4°C, 52% of the biebrich scarlet was retained by the contracting disc and in the second experiment with a shorter time at 4°C, 36% of the biebrich scarlet was retained by the disc. An extinction coefficient for biebrich scarlet of 1.77 x 10^4 L/mole-cm at 505 nm was used.

After three brief rinsings at 45°C with 0.5 mL each of phosphate buffered saline (PBS), 0.15 N NaCl and 0.01 M phosphate at pH 7.4, the 24 h release into a 1 mL aliquot of PBS at 37°C was followed daily for eleven days. The release data were obtained using both the Pye-Unicam 8610 and the AVIV 14 DS spectrophotometers.

Results

Effect of Chemical Clocks on T_t, the Temperature of the Inverse Temperature Transition. The data of Figure 3 demonstrate the concept and potential utilization of chemical clocks. Polymer I, poly(GVGVP), exhibits a constant value for T_t over the three-week period. This polymer is stable in 0.2 M phosphate, pH 7.0 to 7.5. Polymers II and III, poly[0.85(GVGVP), 0.15(GVGcVP)] and poly[0.90(VPGVG), 0.10(VPG$_c$VG)], contain the glycolic acid residue (G_c) and are capable of hydrolytic cleavage with the rupturing of a backbone bond and production of a carboxylate. Both the decrease in chain length and the formation of the carboxylate raise the value of T_t. The value of T_t for these polymers appears to change at different rates in the 6 to 14 day range. The slopes are 0.332°C/day for polymer II and 0.223°C/day for polymer III. When correction is made for the mole fraction differences of the G_c-containing pentamers, however, that is 0.332°C/0.15 mole fraction pentamer-day and 0.223°C/0.1 mole fraction pentamer-day, the values are seen to be identical, 2.2°C/pentamer-day. Therefore, there appears to be no difference in stability for G_c in the VG$_c$V sequence or in the PG$_c$V sequence. The slope for the same period of time, 6 to 14 days, for polymer IV, poly[0.94(VPGVG), 0.06(VPG$_c$NG)] is 0.39°C/day which when corrected for mole fraction, i.e., 0.39/0.06, becomes 6.5°C/pentamer-day. Thus, the combination of both G_c and N results in a three-fold greater slope per pentamer. As will be discussed below, the critical issue becomes the time required for T_t to reach the operating temperature.

Diffusional Release of Biebrich Scarlet from X^{20}-poly(GVGVP). As shown in Figure 3, the value of T_t for polymer I, poly(GVGVP) does not change in phosphate

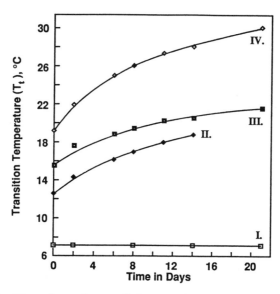

Figure 3. Time dependence of the transition temperature, T_t, for the onset of folding and assembly, which is complete by $T_t +15°C$. Consequently, as the value of T_t rises above 22°C toward 37°C due to backbone esters and/or side chain carboxamides breaking down to form carboxylates, the polymers will dissolve, or the cross-linked matrices would swell, causing a timed release of contents in proportion to the rate at which T_t approaches 37°C.

buffer. Accordingly, release from X^{20}-poly(GVGVP) may be considered entirely diffusional with neither degradation nor change in the degree of swelling altering the release rate.

As noted in the **Methods** section, it is possible to load the X^{20}-poly(GVGVP) matrix in the form of a thin disc with biebrich scarlet simply by lowering the temperature in the presence of a 5 to 6-fold volume of a 0.1 M solution. The release of the loaded contracted matrix at 37° C occurs over a period of more than ten days (Figure 4). Initial release rates are of the order of several hundred nanomoles per day. In the period of 5 to 11 days release rates go from 50 nanomoles/day to 15 nanomoles/day. Thus, even with this thin disc a reasonably narrow range of release rates can be sustained for almost a week. Some differences have been observed depending on the initial equilibration time for swell-doping at low temperature. While variations in the initial washes may be responsible in part, the sample having had the longer equilibration time initially exhibited lower release rates but then higher release rates after five days.

Qualitatively, similar data have been obtained using methylene blue as the drug release model. This study was complicated, however, by an aggregation dependent absorption at the concentration range of interest.

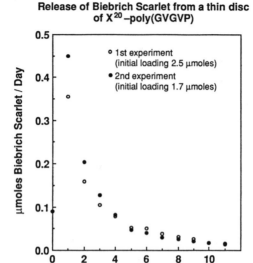

Figure 4. Release profiles for biebrich scarlet from a thin disc of 0.035 mm thickness and volume of 8.2µL. This release is a limiting case due to the absence of reservoir depth, and the absence ester and carboxamide chemical clocks to sustain release. The concentration of biebrich scarlet within the disc can be as high as 0.3M.

Discussion

As illustrated by Figure 4, the property of an inverse temperature transition can be used to load an elastomeric polypeptide matrix with drug. The concept of lowering the temperature to achieve swell-doping and of contracting to expel excess water while retaining drug is clearly demonstrated. In the case of biebrich scarlet, the loading process was achieved at a concentration of 0.3 M in the contracted matrix or approximately 300 μmoles in one cm^3 (one mL) of volume. This is approximately one biebrich scarlet molecule for thee pentamers or one per turn of the ß-spiral. It is apparent that the X^{20}-poly(GVGVP) matrix itself can function as a substantial reservoir for drugs having the chemical composition of biebrich scarlet.

In testing the potential of X^{20}-poly(GVGVP) to achieve a sustained release, a thin monolith was used as the limiting case. In the present report, a volume of 8.2 μL was used within the shape of a thin (0.035 cm thick) disc. Given this challenge, it is encouraging to see the release profile in Figure 4. For thicker monoliths it could be expected that a desired release range, controlled by the concentration in the loaded matrix, be sustained for week periods. The thinness of the present disc is shown in Figure 5, where a stack of 122 discs are required to produce a volume of one mL.

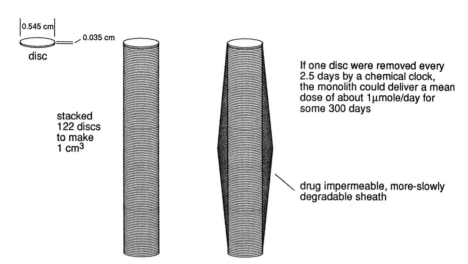

Figure 5. A stack of 122 discs of the size and shape used in Figure 4 are required to achieve a volume of one cm³. With such a depth, with limited release from the sides and with chemical clocks to control swelling, release and degradation from the ends, there is the potential for a sustained release of a high concentration for much of a year.

With the introduction of the matrix ability to degrade, as would be the case for polymers **II** and **III**, the release ranges could be made narrower and extended over longer periods of time. With the introduction of transductional release, using any of the free energy inputs of Figures 1 and 2 or simply using chemical clocks such as carboxamides breaking down to carboxylates at the matrix-milieu interface to control the rate of matrix surface swelling, one can expect to gain even finer control of the drug amount released over even longer periods of time.

Variations in the matrix composition, based on the knowledge of the hydrophobicity scale *(14)*, can be used to change affinities for different drugs or for a given drug. Designs providing oscillatory release over substantial periods of time can also be developed.

The Use of the T_t Change Rate to Effect Transductional Release by Swelling. As long as the value of T_t is 15°C below the physiological temperature, release of the drug would be limited to diffusion from the contracted matrix. However, as T_t increases from 15°C below 37°C to physiological temperature, the affected surface layer of matrix will undergo a swelling to about ten times of its original volume, thereby facilitating diffusional release. This transductional change would not only enhance the release rate but would result in a matrix more susceptible to proteolytic degradation. Should the chemical clock include the glycolic acid residues, there would be the simultaneous hydrolytic cleavage to cause chain rupture. In the absence of such transductional effects, the contracted X^{20}-poly(GVGVP) matrix appeared to remain intact and transparent for many months in the peritoneal cavity of rats *(5)*, and for two months beneath the conjunctiva of the eye in rabbits *(4)*.

If a therapeutic dose for local release of biebrich scarlet to stimulate wound healing (15 to 50 nmoles/day) is to be maintained for a week, then the swell-doped disc of the present study, pretreated by a four-day washout, will suffice. If higher doses are sought over a longer period, then a larger monolith will be used. Shown in Figure 5 is an example of a monolith equivalent to a stack of 122 discs of the present study. If the introduced chemical clock was designed so that an equivalent of one disc was removed every 2.5 days, then the monolith could deliver an average dose as high as one μmole/day for a period of 300 days.

Acknowledgments.   This   work   was   supported   in   part   by   contracts N-00014-90-C-0265 from the Naval Medical Research and Development Command, Office of Naval Research; N-00014-89-J-1970 from the Department of the Navy, Office of Naval Research; and I-R43HL49705-01 from the National Heart, Lung, and Blood Institute, National Institutes of Health.   The authors are pleased to acknowledge Richard Knight of the Auburn University Nuclear Science Center for carrying out the gamma-irradiation cross-linking.

Literature Cited

1. Urry, D. W. *Prog. Biophys. Molec. Biol.* **1992**, 57, 23-57.
2. Urry, D. W. *Angew. Chem. Int. Ed. Engl.* **1993**, 32, 819-841.
3. Urry, D. W.; Parker, T. M.; Reid, M. C.; Gowda, D. C. *J. Bioactive Compatible Polym.* **1991**, 6, 263-282.
4. Elsas, F. J.; Gowda, D. C.; Urry, D. W. *J. Pediatr. Ophthalmol. Strabismus* **1992**, 29, 284-286.
5. Hoban, L. D.; Pierce, M.; Quance, J.; Hayward, I.; McKee, A.; Gowda, D. C.; Urry, D. W.; Williams, T. *J. Surgical Res.* **1993**, in press.
6. Urry, D. W.; Trapane, T. L.; Prasad, K. U. *Biopolymers* **1985**, 24, 2345-2356.
7. Urry, D. W.; Trapane, T. L.; Iqbal, M.; Venkatachalam, C. M.; Prasad, K. U., *Biochemistry* **1985**, 24, 5182-5189.
8. Urry, D. W.; Trapane, T. L.; McMichens, R. B.; Iqbal, M.; Harris, R. D.; Prasad, K. U. *Biopolymers* **1986**, 25, S209-S228.
9. Urry, D. W. In *Cosmetic and Pharmaceutical Applications of Polymers;* Gebelein, C. G; Cheng, T. C.; Yang, V. C., Eds.; Plenum Press, New York, 1991; 181-192.
10. Urry, D. W. *Am. Chem. Soc., Div. Polym. Mater.: Sci. and Eng.* **1990**, 63, 329-336.
11. Prasad, K. U.; Iqbal, M. A.; Urry, D. W. *Int. J. Pept. and Protein Res.* **1985**, 25, 408-413.
12. Urry, D. W.; Prasad, K. U. In *Biocompatibility of Tissue Analogues*; Williams, D. F., Ed.; CRC Press, Inc.: Boca Raton, Florida, 1985; pp 89-116
13. Nicol, A.; Gowda, D. C.; Urry, D. W. *J. Biomed. Mater. Res.* **1992**, 26, 393-413.
14. Urry, D. W.; Gowda, D. C.; Parker, T. M.; Luan, C-H.; Reid, M. C.; Harris, C. M.; Pattanaik, A.; Harris, R. D., *Biopolymers* **1992**, 32, 1243-1250.

RECEIVED September 20, 1993

Chapter 3

Sequence-Selective Binding of DNA by Oligopeptides as a Novel Approach to Drug Design

Moses Lee and Clint Walker

Department of Chemistry, Furman University, Greenville, SC 29613

Recent interest in the development of sequence-specific DNA binding agents as chemotherapeutic agents, structural probes of DNA, and artificial restriction enzymes has led to the development of minor groove DNA binding agents that recognize selective base sequences. A rational approach to the development of low molecular weight oligoimidazolecarboxamide and polymethylene linked analogs of distamycin targeted for long guanine-cytosine (GC) rich sequences of DNA are described.

Many clinically useful anticancer agents exert their bioactivity by interacting with DNA. Severe toxic side effects of these agents have been attributed to their limited sequence selectivity. Consequently, this has led to the development of sequence specific DNA binding agents for use as chemotherapeutic agents, structural probes of DNA, and artificial restriction enzymes (1,2). Certain regions of the human genome, including the regulatory regions of oncogenes (e.g., c-Ha-**ras**) (3) and the 3-Kb units of the Epstein Barr virus genome, have unusually high guanine-cytosine (GC) contents (>80%) (4). These observations suggest that GC runs are possible targets for the development of DNA sequence binding ligands as potential anticancer and gene control agents. When the human genome is the target for sequence selective agents, the distinguishable sequence must have a binding site size of 15 to 16 base pairs (1c).

Our approach to the design of agents that can recognize specific DNA sequences uses the naturally occurring minor groove and AT sequence specific oligopeptides netropsin and distamycin as models. These oligopeptides display antibiotic and antiviral properties (5) by blocking the template function of DNA. They are known to bind specifically to 4 and 5 contiguous adenine-thymine (AT) base pairs, respectively, in the minor groove of double helical B-DNA (1,2,5). The sequence specificity is the result of hydrogen bonding, electrostatic attraction, and van der Waals interactions (3). The concave hydrogen of the pyrroles in netropsin prevents it from binding to a GC base pair, thus forcing the recognition of an AT

0097–6156/94/0545–0029$08.00/0

base pair by default *(6,7)*. Thus substitution of a heterocycle that provides space and a hydrogen bond acceptor for G(2)-NH$_2$ in the place of pyrrole rings should alter the strict AT binding preference to permit GC recognition *(6,7)*. The imidazole containing analogs of netropsin were shown to exhibit specificity for certain GC rich sequences *(8,9)*. In our opinion, the imidazole containing analogs of distamycin are well suited for the development of agents that can recognize long GC rich sequences of DNA.

Dervan's group has shown that oligopeptides with four, five, and six N-methylpyrrole-carboxamide units bind to AT-rich sequences of five, six, and seven contiguous base pairs, respectively *(1c)*. The general rule is that **n** amides afford binding site sizes of **n + 1** base pairs *(1c)*. The main problem associated with the development of polyheterocyclic analogs to read long DNA sequences concerns the relationship between the repeat distance of a nucleotide unit of DNA and the hydrogen bond and van der Waals contacts generated by an N-methylpyrrole peptide, *i.e.*, the "phasing" problem. According to Goodsell and Dickerson, binding of netropsin isohelically to DNA is "out of phase" by 1 Å per monomer (pyrroleamide) step *(10)*. Hence, simply increasing the length of the N-methylpyrrole oligopeptide to gain additional degrees of specificity will result in a severe mismatch between the peptide and DNA and will lower its DNA binding constant *(1c,11)*. However, there is an optimal number of heterocyclic units that these analogs can have and still effectively bind to DNA. In an attempt to overcome some of these "phasing" problems, linked pyrrole-containing analogs were synthesized with varying lengths of polymethylene chains. They were shown to bind to long (up to 10 base pairs) (AT)$_n$ sequences *(12-14)*.

Our approach to the development of agents that can recognize long GC rich sequences requires, first, the design of optimal oligoimidazolecarboxamido analogs of distamycin as the "reading frame" for the longer agents. In the second part, the "optimal" reading frames will be linked with a flexible polymethylene tether. Several criteria to be met in the development of suitable linkers are: correct length, flexibility, shape, and lipophilicity *(13)*. In these compounds, dimethylamino groups are used as the cationic groups since they are synthetically more accessible, and they have sequence selectivity similar to the amidines.

Sequence Selective Binding of Oligoimidazolecarboxamides

DNA Binding Studies. The oligoimidazole-containing analogs *(1-4)* of distamycin have been obtained (Figure 1). The apparent binding constants, K$_{app}$, of these

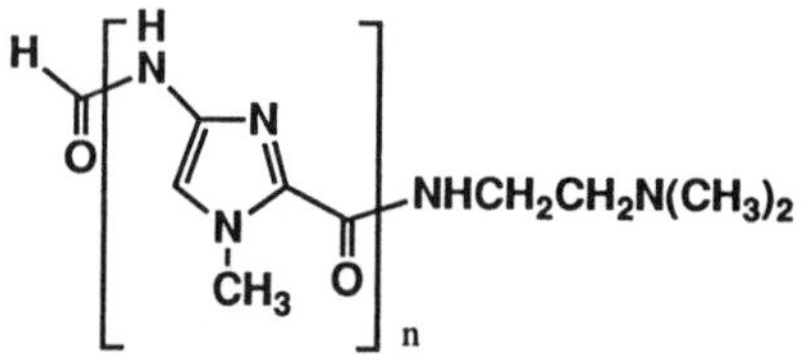

Figure 1. Structure of the oligoimidazole analogs **1-4**.

compounds *(1-4)* and distamycin A to calf thymus DNA, T4 coliphage DNA, poly(dA.dT), and poly(dG.dC) were determined from the ethidium binding assay *(15)* and are presented in Table 1. These data demonstrate that the oligoimidazole analogs can bind to the DNA's studied. The apparent binding constants, obtained at room temperature, are meant to compare relative values rather than absolute values. The values of K_{app} for 1 and 2 are lower than that of the distamycin, presumably due to the lower number of amide moieties in 1 and 2 (2 and 3, respectively versus 4) and van der Waals contacts compared to distamycin. The K_{app} values of the triimidazole analog 3 are comparable to distamycin for calf thymus and T4 coliphage DNA. All the above compounds bind stronger to poly(dG.dC) than poly(dA.dT) in contrast to distamycin. This indicates that changing the pyrroles in netropsin to imidazoles increases the acceptance of these compounds for GC base pairs. However, as the number of imidazole units was increased from three to four, a slight decrease in the apparent DNA binding constants was observed suggesting that the optimum number of heterocyclic units of these analogs for DNA sequence selective recognition is three.

The apparent binding constants for T4 coliphage DNA give an indication of the groove selectivity of these compounds. The major groove of T4 coliphage DNA is blocked by α-glycosylation of the 5-hydroxymethylcytidine residues; therefore, the only place available for non-intercalating agents to bind is the minor groove *(16)*.

ΔTM Studies. The relative effect on the melting temperature of the DNA in the drug/DNA complex is compared to that of the free DNA. The results are reported as ΔTM. For our experimental conditions, the ΔTM of drug:calf thymus DNA complex at **r'** of 0.25 (**r'** = number of moles of drug to the number of moles of DNA base pairs) observed for this series of oligoimidazole analogs 1, 2, 3, and 4 were 4.3°, 2.9°, 4.5°, and 4.0°C, respectively. The melting temperature of free calf

Table 1. Association Constants (K_{app} ± 0.02 x 105 M^{-1}) of Compounds with Polynucleotides

Compound	Calf thymus	T4	(dA.dT)n	(dG.dC)n
EtBr	100 *(15a)*	100 *(15b)*	91 *(15b)*	99 *(15b)*
Dist	7.74	6.50	348	2.03
1	0.25	-	-	0.02
2	0.32	0.40	0.17	0.23
3	7.74	6.67	5.94	6.13
4	3.43	3.33	4.84	5.34

thymus was 56.8°C. The largest ΔTM value was found for the triimidazole ligand **3**. An increase in the number of imidazole units to four, **4**, caused a slight decrease in the ΔTM value. This data further suggest that the optimal number of imidazole units for DNA binding is three.

Circular Dichroism Studies. The conformation of DNA can be readily determined by CD experiments. For B-DNA, characteristic positive and negative Cotton effects are observed at 260 and 245 nm, respectively (Figure 2a). However, interaction of molecules with the DNA can cause changes in the CD spectrum, such as the appearance of ligand-induced bands and/or alteration in the original CD spectrum (*1b*).

Titration of the monoimidazole analog **1** to poly(dG.dC) and poly(dA.dT) showed no change in the CD spectra, suggesting that the ligand has minimal interaction with the DNA's, which is consistent with its low K_{app} values. Further results show that compounds **2-4** bind to the studied DNA's, as indicated by the appearance of ligand-induced CD bands at about 300-340 nm, presumably due to the UV absorption π to π^* transition of the ligand in the ligand:DNA complex. The appearance of a ligand-induced CD band is clear evidence of the interaction of the molecules with the DNA because these compounds do not exhibit any CD spectra by themselves. Specifically, at equimolar drug concentrations, ligand **2** gave a stronger positive Cotton effect at 295 and 315 nm for poly(dG.dC) and calf thymus DNA than for poly(dA.dT) at 310 nm, thus indicating a preference for GC sequences. In each of these titration experiments, a clear isodichroic point at 280-295 nm was observed. This suggests a single mode of interaction between the drug and the DNA, *i.e.*, minor groove binding, assuming that the drug exists in two forms, free and bound. This observation is in agreement with the data obtained from the ethidium displacement assay which showed that **2** binds to coliphage T4 DNA.

Titration of the triimidazole analog **3** to poly(dA.dT) produced a negative Cotton effect at 308 nm (negligible ellipticity at $r' = 0.2$) and an isodichroic point at 289 nm (Figure 2a). However, titration of **3** to poly(dG.dC) gave rise to a significantly more intense positive Cotton effect band at 332 nm (1.8 mdeg, $r' = 0.25$), a negative band at 300 nm (negligible ellipticity at $r' = 0.25$), and an isodichroic point at 310 nm (Figure 2b). Titration of this compound to calf thymus DNA at low drug concentrations ($r' < 0.6$) also produced a strong positive band at 332 nm (5 mdeg, $r' = 0.2$), a negative band at 300 nm (1.5 mdeg, $r' = 0.2$), and an isodichroic point at 305 nm (Figure 2c). However, at higher drug concentrations ($r' > 0.6$) the negative band at 305 nm intensified and the curve shifted off the isodichroic point indicating changes in the interaction mode(s). The data on the ligand-induced ellipticity (Figure 3) strongly suggest that compound **3** binds stronger to poly(dG.dC) and calf thymus DNA than poly(dA.dT).

The interactions of the tetraimidazole analog **4** with the above three polydeoxyribonucleotides, at equimolar concentrations, were also studied by CD studies. Addition of **4** to poly(dA.dT) produced a weak band at ≈320 nm (0.4 mdeg, $r' = 0.2$), a slight decrease in the positive band at 260 nm, and an isodichroic point

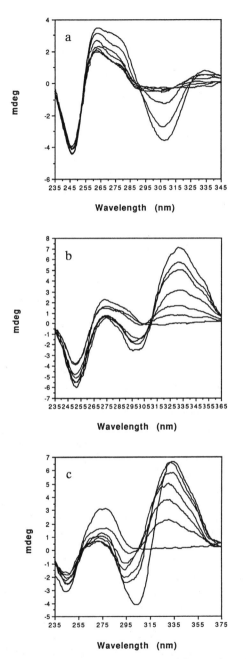

Figure 2. Titration of the triimidazole analog **3** to (a) poly(dA.dT), (b) poly(dG.dC), (c) calf thymus DNA. The plots correspond to **r'** values of 0, 0.05, 0.10, 0.20, 0.40, 0.60, 0.80, and 1.0 for poly(dA.dT), and calf thymus DNA. For poly(dG.dC), the **r'** are 0, 0.12, 0.25, 0.35, 0.5 and 0.6.

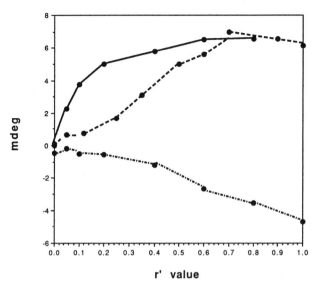

Figure 3. Variation of ellipticity (mdeg) of the ligand-induced bands (300-340 nm) of various DNA's as a function of **r'** for compound 3:poly(dG.dC) (---), -calf thymus DNA (—), -poly(dA.dT) (-.-.-).

at 280 nm. When it was titrated into poly(dG.dC), a negative band at 315 nm (0.6 mdeg, **r'** = 0.25) was seen which was bigger than that seen for poly(dA.dT). Titration of **4** into calf thymus DNA also produced a weak negative ligand-induced band at 320 nm, thus suggesting that it binds weaker to DNA compared to **3**.

In the above CD experiments, the presence of the negative and positive bands of the DNA at ≈245 and ≈270 nm, respectively, suggests that the conformation of the DNA in the ligand:DNA complexes remained in the B-form (Figure 2). Data from CD titration studies also provide information on the binding site size of the ligands at saturation concentration from plots of induced ellipticity (mdeg) at 290-340 nm versus the values of **r'**. Such plots for compound **2** and calf thymus DNA showed saturation at a **r'** value of 0.25, suggesting that each drug molecule spans a binding site size of four base pairs. However, the plots of the induced ellipticity for the titration of **3** to calf thymus DNA gave a binding stoichiometry of **r'** ≈ 0.2, indicating that the drug binds to about five DNA base pairs (Figure 3). These results are in agreement with Dervan's **n + 1** rule, wherein compounds **2** and **3** which have three and four amide groups would bind to four and five contiguous base pairs, respectively *(1c)*. Titration of **4** into calf thymus DNA produced a binding stoichiometry of **r'** ≈ 0.2 suggesting that the ligand could span a binding site size of five base pairs, albeit at lower affinity than the triimidazole analog **3**.

The results from the CD studies indicate, on the basis of the molar ellipticity of the ligand-induced bands, that the strongest binder is the triimidazole compound **3**, followed by the diimidazole analog **2**, then tetraimidazole **4**, and finally monoimidazole **1**. The decrease in the binding of **4** compared to **3** suggests that the former compound may have a less compatible geometry and shape for interacting isohelically with the minor groove of the DNA. Therefore, in order to further investigate the structural requirements for the interactions of these analogs with DNA, an ^{1}H-NMR study on the 1:1 complex of **2** with the decadeoxyribonucleotide d-[CATGGCCATG]$_2$ *(9)* was undertaken.

Titration and Binding of 2 to d-[C1A2T3G4G5C6C7A8T9G10]$_2$. The results of the titration of **2** to the decadeoxyribonucleotide are shown in Figures 4b-d, together with the assigned reference spectrum of the free DNA (Figure 4a) *(9)*. With increasing proportions of **2**, the 1H-NMR resonances of the decamer show distortion of line shape resulting from signal broadening and ligand-induced chemical shift changes. Some new resonances appear which increase in proportion to the amount of drug added (Figures 4b,c). The changes in the chemical shifts arise from changes in the chemical environment of the oligonucleotide upon formation of the 2:decamer complex, while the line broadening was due to the increase in rotational correlation time from the binding of a small ligand to a macromolecule *(9)*. The spectrum of the 1:1 complex at 25°C (Figure 4c) indicated loss of degeneracy (doubling of resonances) at the sites CH6(6,7) and TH6(9) which showed that the exchange process of the ligand on the DNA was slow on the NMR time scale, and it provides an indication for the location of **2** at these sites on the decamer. As expected, when the temperature was raised, the 1H-NMR signals of the 1:1 complex sharpened indicating that the ligand was exchanging on the DNA (Figure 4d).

Assignments of the Non-Exchangeable Protons and Location of 2 on d-[CATGGCCATG]$_2$. The non-exchangeable 1H-NMR signals of the 1:1 complex were assigned by using a combination of NOE difference, COSY and NOESY methods *(9)*. The NOEs, observed between H8 and H2'/2", H3' and H1' of A(2), with relative intensity AH2'/2" >> AH3' ≥ AH1', indicate that the conformation of the 1:1 complex belongs to the B-family *(18)*. This supports the CD data that the ligand binding with DNA only causes minimal conformational changes on the nucleic acid.

The ligand-induced chemical shift changes of the aromatic protons for each of the bases in the oligomer [$\Delta\delta$ (δ 1:1 complex - δ free DNA) 0, -0.23, -0.04, -0.10, +0.06, +0.14, +0.11, -0.31, -0.31, -0.06 ppm for purine-H8, TH6 and CH5 of CATGGCCATG, respectively] showed that nucleotides C(6) to A(9) were most influenced by the binding of **2**. The location of **2** on this sequence was further corroborated by the results from NOE difference studies. Saturation of the -N(CH3)$_2$ group at 3.02 ppm of **2** gave NOEs at ≈7.54 ppm and ≈7.86 ppm indicative of its close proximity to AH2(2,8) (data not shown). Furthermore, saturation of the signal at 7.86 ppm gave NOEs to CC$\underline{H}_2$NMe$_2$, and irradiation at 7.54 ppm produced a strong NOE to the -N(C$\underline{H}_3$)$_2$ group of **2**. Accordingly, the signals at 7.54, 7.56 (exchanging signals), and 7.86 ppm were ascribed to AH2(2 and 8), respectively. These data illustrate that the ligand is bound in the minor

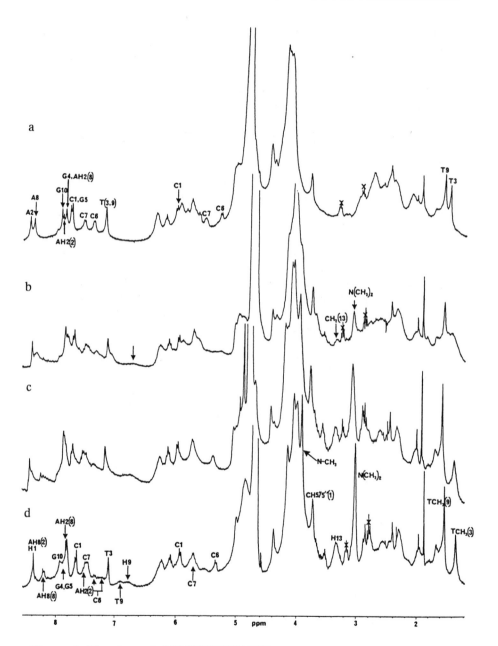

Figure 4. Titration of d-[CATGGCCATG]$_2$ with ligand **2**. Selected ^{1}H-NMR assignments are indicated in the spectrum. Part **a** is the spectrum of the free DNA at 25 °C. Parts **b** and **c** correspond to the addition of 0.5 and 1.0 mole equivalent of **2**, respectively, at 25 °C. Part **d** represents the spectrum of the 1:1 ligand **2**/DNA complex at 37 °C.

groove of the C_6CAT_9 and the dimethylamino moiety at the C-terminus of **2** is oriented to T(9) while the imidazole moieties are situated in the GC region. This investigation also shows that the second methylene group on the C-terminus is in a close proximity to AH2(2) which is responsible for reading the 3' AT site. These data are consistent with the previous reports *(9)* that it is the van der Waals interactions that prevent the binding of the C-terminus to a GC site thus forcing it to read AT sites. A model for the 1:1 complex of **2** and 5'-CCAT-3' of the decamer is depicted in Figure 5.

MPE Footprinting. Compounds **1-4** were initially screened over a large dose range on a large fragment (Hind III/EcoR 1) of pBR322 DNA (data not shown). The mono- and tetraimidazole analogs **1** and **4** gave poor footprints. The GC recognition of this compound was confirmed by densitometric analysis and, in particular, the binding of **2** to the sequence 5'-CCGT-3'. Compound **3** produced clear footprints although at different sites from **2**.

In order to analyze in more detail the binding of **2** and **3**, a 275 base pair

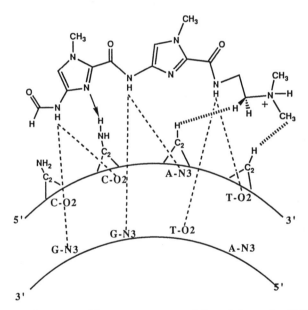

Figure 5. Molecular recognition components of ligand **2** on the 5'-CCAT sequence of the decadeoxyribonucleotide. Dotted lines represent hydrogen bonding between the amide NH groups on cytosine-O2, Guanine-N3, adenine-N3 and thymine-02. The thick arrow indicates the new hydrogen bond between G-2-NH$_2$ and the imidazole-N3(N2). The NOEs observed between AH$_2$(2) and -N(CH$_3$)$_2$ and H13 of **2** and AH2(8) and -N(CH$_3$)$_2$ of **2** are denoted by the dashed lines.

GC rich fragment of pBR322 (Bam H1/Sal 1) was isolated and data were obtained from both strands. Distinct footprints are observed for both compounds, and it is clear that they do not recognize the same sequences even though they differ only by one imidazole unit. These results clearly show that small changes in the molecules can produce significant changes in their sequence recognition.

Using electrophoresis runs of differing times, it was possible to resolve an overlapping region of 150 base pairs from both strands. A detailed densitometric analysis was performed in this region; the data are summarized in Figure 6. Both the recognition of GC rich sequences and the differences between **2** (Figure 6a) and **3** (Figure 6b) are clearly evident. For compound **2**, a general preference for 5'-(G.C)$_3$(A.T) is observed, whereas the strongest sites for **3** are within two occurrences of the sequence 5'-TCGGGCT which are not recognized by compound **2** (see Figure 6).

Molecular Modeling Studies. In order to follow the minor groove of DNA, the ligand must adopt a conformation isohelical to the DNA. Force field calculations predicted the minimum energy conformation for the imidazole analogs **1-4** in which the amide bonds are fixed in the s-Z conformations. The conformation preference is caused by a stereoelectronic factor which cannot be determined by this calculation *(19)*. The energy minimized conformations of **1-4** show that each molecule adopts a crescent shaped conformation which matches the convex surface of the minor groove of DNA. The planes of the imidazoles are coplanar, thus providing the necessary shape to fit snugly into the minor groove of the DNA and permitting the amide protons and the imidazole-N$_3$ group to form hydrogen bonds with the electron rich sites, such as guanine-N$_3$, -2-NH$_2$, adenine-N$_3$, thymine-O$_2$, and cytosine-O$_2$.

An ideal distance of the helical axis to the amide hydrogen of the ligand, which is isohelically bound to DNA, is estimated to be 4.5 Å. For an ideal B-DNA (h = 3.38 Å, t = 36°, π = 47.1°) with this radius (r = 4.5 Å) the repeating unit of the perfectly located ligand hydrogen, capable of forming hydrogen bonds, can be calculated to be 4.4 Å *(10,20)*. In the energy minimized conformation of **1-4**, the distances between the amido-NH groups are from 4.5 Å to 4.9 Å, thus indicating that the formation of hydrogen bonds between consecutive base pairs and the ligand should be possible.

Further molecular modeling studies included consideration of the curvature of the imidazole analogs to fit isohelically into the convex surface in the minor groove of DNA. It has been shown that the curvature of netropsin and distamycin is 15 Å which matches well with the curvature of the minor groove of AAAA (19 Å) thus explaining their strong binding to DNA *(21)*. In our studies, the curvature of the tetranucleotide d-[G$_4$.C$_4$], drawn in the B-conformation using the DNA/RNA builder program and displayed on the Tektronix-CAChe system, was determined to be 19 ± 1 Å, using the G-2-NH$_2$ groups as points on the circle. From the energy minimized conformations of **2-4**, their curvature were measured to be 15, 17 and 13 ± 1 Å, respectively, using the carbon of the formyl moiety and the N-1 atom of the imidazoles. The increase in the curvature of the tetraimidazole analog may explain the decrease in its binding constant to DNA.

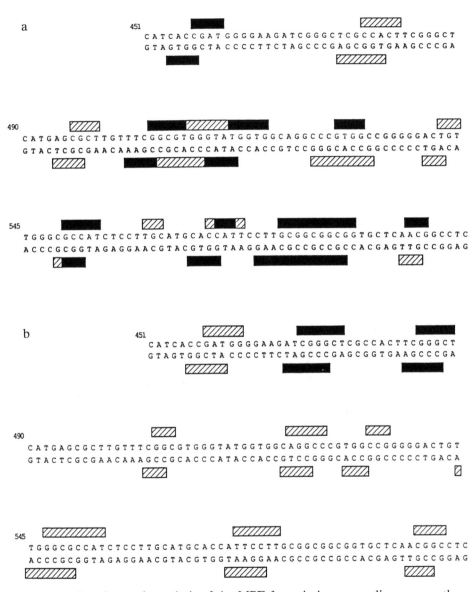

Figure 6. Densitometric analysis of the MPE footprinting autoradiograms on the Bam H1/Sal 1 fragment of pBR322 DNA. Solid and shaded boxes represent strong and weak footprint sites. Top boxes are for the Bam H1 labeled, and the bottom boxes are for the Sal 1 labelled, DNA. Parts **a** and **b** are for compounds **2** and **3**, respectively.

In order to understand the implications of increased curvature and reduced binding constant, the energy minimized conformers of **1-4** were docked into the minor groove of the following B-oligomers: 5'-C<u>CAT</u>, 5'-C<u>CCAT</u>, 5'-C<u>CCCAT</u>, and 5'-C<u>CCCCAT</u>, respectively. In these complexes, the ligands are located on the underlined sequence with the dimethylamino group residing over the T residue as construed from the ^{1}H-NMR data obtained. The models, while not energy minimized, show that analog **3** fits snugly into the minor groove (see Figure 7a), and all the amido-NH groups form bifurcated hydrogen bonds with the electron-rich sites (distances between the amido-N<u>H</u> groups and the purine-N3 and pyrimidine-O2 are between 2.6 - 3.2 Å). However, due to the increase in the curvature of **4**, it is difficult to fit properly in the minor groove. If the terminal amido-NH groups were positioned for hydrogen bonding with the electron rich sites in the minor groove, then the central amido-NH groups would be positioned too far (5-6 Å) for hydrogen bonding with the DNA (Figure 7b) and thus destabilize the ligand:DNA complex. This could explain the lower DNA affinity for the tetraimidazole analog **7** compared to the "shorter" triimidazole analog.

These studies clearly indicate that the length of three imidazole units is optimal for this class of compounds, and any further increase in the number of

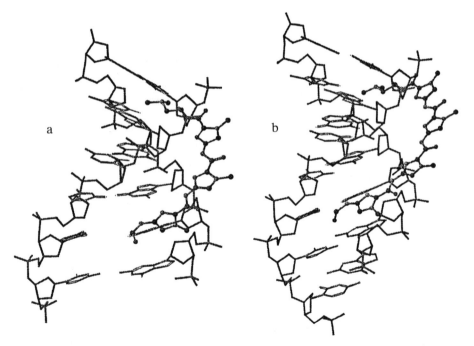

Figure 7. The energy minimized conformer of the imidazole analogs **3** and **4** were docked into the minor groove of the underlined sequence of the B-oligonucleotides (a) 5'-C<u>CCCAT</u> and (b) 5'-C<u>CCCCAT</u>, respectively, with the dimethylamino moiety located over the T residue. The 1:1 complexes were not energy minimized.

heterocycles to target for longer DNA sequences is not feasible due the 'phasing' problem. Therefore, future developments of such imidazole containing analogs of distamycin for long GC rich sequences would have to include the use of linkers or develop heterocycles with smaller repeating units.

Sequence Selective Binding of N-to-N Oligomethylene Linked Analogs

DNA Binding Studies. One approach to the development of molecules that could bind to long DNA sequences involves the preparation of analogs **5(a-f)** of distamycin (see Figure 8) in which the N-termini of the "reading frames" are linked with an oligomethylene tether. The relative DNA binding constants (K_{app}) of these compounds are given in Table 2. The compounds' significant binding constants to T4 DNA demonstrate that they bind to the minor groove. The GC sequence binding selectivity of these compounds is inferred by comparing their binding constants to poly(dG.dC) and poly(dA.dT) to those of distamycin (see Table 1).

These K_{app} values show two maxima at **n** = 2 (number of methylene groups) (**5b**) and **n** = 6 (**5f**) for each of the four DNA's examined. This suggests that these

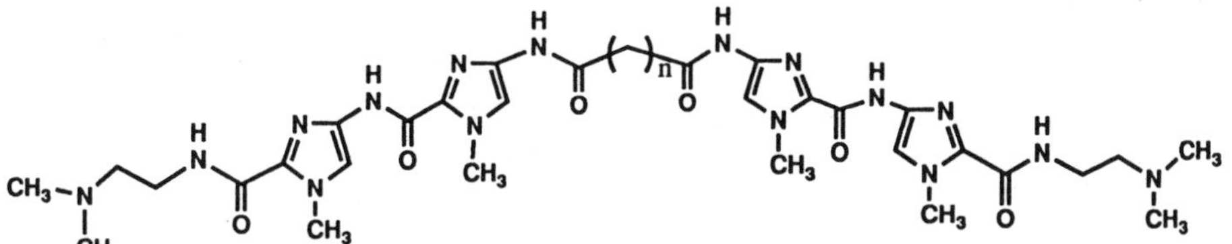

5a-f, n=1 to 6, respectively

Figure 8. Structures of the linked analogs **5a-f**.

Table 2. Apparent DNA Binding Constants ($\pm 0.1 \times 10^5$ M^{-1})

	Calf thymus	T_4	(dA.dT)n	(dG.dC)n
5a	1.5	2.0	1.6	1.8
5b	5.8	9.0	3.2	5.4
5c	1.1	1.1	0.9	0.5
5d	3.7	4.0	2.5	4.3
5e	1.1	1.3	1.1	1.0
5f	8.7	11	8.6	6.6

two compounds must be able to adopt optimal conformations that would bind isohelically to the minor groove with minimal energy sacrifice compared to the others.

Another indicator of GC sequence preference by analogs **5b, d** and **f** is the larger positive Cotton effect bands of the ligand- induced CD band at 295-350 nm for poly(dG.dC) than for poly(dA.dT) at comparable drug concentrations. Results from these studies also show that the ligand-induced ellipticity for poly(dG.dC) at **r'** = 0.2 was strongest for **5f** (6.1 mdeg), followed by **5b** (5.8 mdeg) than **5d** (3.1 mdeg). This suggests that the hexamethylene-linked compound binds more strongly to the DNA's studied than the other compounds. The CD titration spectra also show clear isodichroic point(s) which indicate a single mode for binding the ligand to the minor groove. Based on the plots of ligand-induced Cotton effects against **r'** at low concentration, compounds **5b, d, f** bind to approximately 7-10 base pairs on poly(dG.dC) and poly(dA.dT) in a bidentate fashion (point of inflection $r' \approx$ 0.05-0.15). However, at higher drug concentrations they bind to 3-4 bases in a monodentate fashion ($r \approx 0.3$) *(14)*. Further examination on the binding of **5f** to poly(dG.dC) (Figure 9a) reveals that at low **r'** values (<0.074), the intensities of the negative and positive bands at 253 and 285 nm, respectively, decrease compared to the free DNA. This is indicative of the ligand binding to the DNA. However, as the **r'** values increase, the positive band shifted to 293 nm, and both bands dramatically intensified. For poly(dA.dT) at a low **r'** of 0.15 (Figure 9b), the ellipticities of the two positive bands at 291 and 322 nm were 0.8 and 1.2 mdeg, respectively. However, at a higher **r'** value of 0.3, the shorter wavelength band intensified dramatically (4.0 and 2.7 mdeg, respectively). These data suggest that the ligand could bind to the DNA in at least two different modes, wherein at low concentrations, they could bind in a bidentate fashion by spanning 7-8 base pairs.

The ability of **5b** to bind to DNA in a bidentate fashion was corroborated by ^{1}H-NMR studies of the 1:1 complex with d-[CATGGCCATG]$_2$. The clearly defined signals for the base protons suggest that, on the NMR time scale, the 1:1 complex exists on average in one form in solution. As the temperature was raised, the signals sharpened indicating that the ligand is changing equivalent binding sites on the DNA. The signals of the 1:1 complex were also assigned on the basis of COSY, NOE and T$_1$ studies. In these studies, NOEs were observed between AH2(2,8) and the dimethyl groups of **5b** suggesting that the ligand is located in the minor groove on the sequence A$_2$TGGCCAT$_9$, with the methyl groups residing on the A(2) and T(9). The structure of this complex was supported by the ligand-induced chemical shifts measurements in which the base protons of A$_2$TGGCCAT$_9$ showed largest $\Delta \delta$ values. The NOE studies also revealed that the binding of **5b** to the oligomer only caused minor changes in the conformation of the nucleic acid backbone, and it remains in the B-form.

The flexibility of the C-C bonds on the oligomethylene linker was demonstrated by MM2 conformational geometry analyses in which the local minima and the global minimum conformers were obtained. The strain energies between the conformers were close to the low energy barriers between them. A two-dimensional dihedral angle driven MM2 energy minimization on **5b** was also

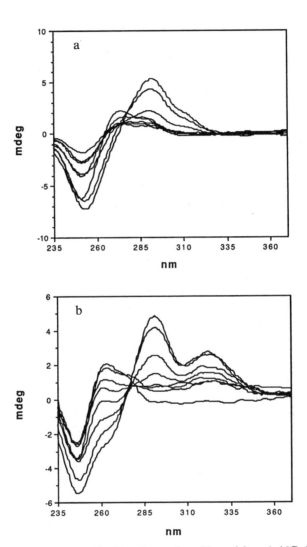

Figure 9. Titration of the triimidazole analog **5f** to (a) poly(dG.dC), (b) poly(dA.dT). The plots correspond to **r'** values of 0, 0.01, 0.024, 0.05, 0.074, 0.10, 0.13, and 0.15 for poly(dG.dC). For poly(dA.dT), the **r'** values are 0, 0.025, 0.05, 0.10, 0.20, 0.40, 0.60, and 0.80.

performed which showed that the rotation of amide bond from <u>anti</u> to <u>syn</u> has a high energy barrier (42.7 kcal), while the rotation of aliphatic bond of the linker has a low energy barrier (4.8 kcal). Therefore, the oligomethylene linked N-to-N bisimidazole analogs **5a-f** can readily achieve the optimal crescent-shape conformation with only a small sacrifice of energy. In this conformation, the amide NH groups are pointing into the concave face of the molecule for hydrogen bonding, with the electron-rich sites (purine-N3 and pyrimidine-O2) on the floor of the minor groove. The binding of lexitropsins and distamycin to the minor groove of DNA is highly exothermic ($\Delta H° < 0$) and thermodynamically favorable ($\Delta G° < 0$) as determined by microcalorimetry studies *(17)*. Therefore, the energy loss from the binding process of **5a-f** to DNA must overcome that gained from changes of the polymethylene linker to attain the optimal DNA binding conformer.

Since the ideal repeat distance '**d**' for the interaction of the amido groups with DNA is 4.4 Å *(10,20)*, the results given in Table 3 suggest that the linker of the optimal DNA binding conformer of **5b** and **5f** could span 2 and 3 base pairs, respectively. The repeat distance '**d**' of the linker of the former compound is slightly greater than ideal; however, the value of '**d**' of the latter molecule is slightly less than two times the ideal number. Since the repeat distance of the imidazole units slightly exceeds the ideal value (4.5-4.9 Å), a shorter linker may be beneficial to compensate for this discrepancy. The other compounds have much less favorable repeat distances for binding to the DNA.

Another structural requirement for the binding of these compounds to the minor groove of DNA is their radius of curvature. The curvature of the minor groove of DNA is ≈19 Å. According to the data given in Table 3, the optimal conformation of compounds **5d-f** have the most compatible curvature for binding to the DNA, followed by **5b**. Due to the unfavorable repeat distance of **5d** and **e**, they interact weaker with DNA than **5f**. Since the radius of curvature of **5b** is much larger than 19 Å, it would not bind to DNA as strongly as **5f** even though its repeat distance is close to ideal. These data support the experimental evidence that the

Table 3. Radius of Curvature and the Repeat Distances of the Optimal Conformers of 5b-f

Compound	Curvature (Å)	Repeat Distance 'd' (Å)
5b	36.4	4.42
5c	27.5	2.90
5d	18.2	9.13
5e	19.8	7.50
5f	20.8	8.70

hexamethylene linked compound binds most strongly to the DNA's studied compared to the other analogs.

In attempt to rationalize the binding of the linked analogs to DNA, the optimal conformer of the linker of **5b** (d = 5.0 Å) was docked into the minor groove of the double stranded d-[GGG.CCC] in the B-conformation. The model shows that the imidazole-linker-imidazole fragment fits snugly into the groove, and the amide-NH groups as well as the imidazole-N3 atoms are in positions capable of hydrogen bonding to the electron-rich sites (G-N3, C-O2) and guanine-2-NH$_2$ groups, respectively.

Encouraged with these results, detailed MPE footprinting studies on compounds **5a-f** are in progress. Preliminary results show that compounds **5b, d** and **f** produce clear footprints, and the others did not. In addition, the hexamethylene linked compound gave strongest footprints at comparable concentrations. The data presented in this chapter shows that the development of linked oligoimidazole analogs is a feasible approach to the design of ligands that can recognize long GC rich sequences of DNA. Furthermore, these linked ligands could serve as "vectors" for the delivery of potent DNA interactive agents to targeted sequences to produce favorable biological responses.

Acknowledgement.   The   authors   acknowledge   the   National   Science Foundation-REU for support.

Literature Cited

1. (a) Krowicki, K.; Lee, M.; Hartley, J.A.; Ward, B.; Kissinger, K.; Skorobogaty, A.; Dabrowiak, J.C.; Lown, J.W. *Structure & Expression* **1988**, 2, 251. (b) Zimmer, C.; Wahnert, U. *Prog. Biophys. Molec. Biol.* **1986**, 47, 31. (c) Dervan, P.B. *Science* **1986**, 234, 464.

2. Lown, J.W. *Anti-Cancer Drug Des.* **1988**, 3, 25.

3. Barbacid, M. Ann. *Rev. Biochem.* **1987**, 56, 7779.

4. Karlin, S. *Proc. Natl. Acad. Sci. USA* **1986**, 83, 6915.

5. Hahn, F.E. In *Antibiotics III. Mechanisms of Action of Antimicrobial and Antitumor Agents.* Corcoran, J.W and Hahn, F.E. Eds. Springer-Verlag, N.Y., **1975**, 79.

6. Kopka, M.L.; Yoon, C.; Goodsell, D.; Pjura P.; Dickerson, R.E. *Proc. Natl. Acad. Sci. USA* **1985**, 82, 1376.

7. Lown, J.W.; *Biochem.* **1986**, 25, 7408.

8. Kissinger, K.; Krowicki, K.; Dabrowiak, J.C.; Lown, J.W. *Biochem.* **1987**, 26, 5590.

9. Lee, M.; Hartley, J.A.; Pon, R.T.; Krowicki K.; Lown, J.W. *Nucleic Acids Res.* **1987**, 16, 665 and references therein.

10. Goodsell, G.; Dickerson, R.E. *J. Med. Chem.* **1986**, 29, 727.

11. Luck, G.; Zimmer, C.; Reinert, K.E.; Arcamone, F. *Nucl. Acid Res.* **1977**, 4, 2655.

12. Griffin, J.H.; Dervan, P.B. *J. Am. Chem. Soc.* **1986**, 108, 5008.

13. Lown, J.W. In *Molec. Basis of Specificity in Nucleic Acid-Drug Interactions.* Pullman, B., Jortner, J., eds. Kluwer Acad. Pub., Netherlands, 1990, p. 103 and references therein.
14. (a) Gursky, G.V., *et al. Synthetic Sequence Specific Ligands* **1983**, Cold Spring Harbor Symposium, Vol. 47, 367. (b) Dasgupta, D.; Parrack, P.; Sasisekharan, V. *Biochem.* **1987**, 26, 6381.
15. (a) Morgan, A.R.; Lee, J. S.; Pulleyblank, D.F.; Murray, N.L.; Evans, D.H. *Nucleic Acids Res.* **1979**, 7, 547. (b) Debart F., *et al. J. Med. Chem.* **1989**, 32, 1074.
16. Lown, J.W. *Acc. Chem. Res.* **1982**, 15, 381.
17. Marky, L.A.; Synder, J.G.; Remeta, D.P.; Breslauer, K.J. In *Biomol. Stereodyn.* Sarma, R.H., ed. Academic Press, New York, 1983, vol. 1, p 487.
18. Gronenborn, A. M., Clore, G.M. *Prog. Nucl. Magn. Reson. Spectrosc.* **1985**, 17, 1.
19. Deslongchamps, P. *Stereoelectronic Effects in Organic Chemistry* **1983**, Pergamon, New York, 101.
20. Kumar, S., Jaseja, M., Zimmerman, J., Yadagiri, B., Pon, R.T., Saspe, A-M., Lown, J.W. *J. Biomol. Str. Dyn.* **1990**, 8, 99.
21. Cory, M.; Tidwell, R.R.; Fairlet, T.A. *J. Med. Chem.* **1992**, 35, 431.

RECEIVED June 18, 1993

Chapter 4

Protein Device for Glucose-Sensitive Release of Insulin

Y. Ito and Y. Imanishi

Department of Polymer Chemistry, Kyoto University,
Kyoto 606–01, Japan

A glucose-sensitive protein device for insulin release was synthesized by coupling insulin to glucose oxidase via a disulfide bond. Glucose oxidation with glucose oxidase produces an electron that cleaves the disulfide bond. Several sophisticated devices are discussed.

The glucose-sensitive insulin-releasing systems that have been developed for the treatment of diabetes include encapsulation of viable pancreatic cells, electromechanical systems (using a glucose sensor and an insulin injection pump), and chemical systems. There are two major types of the chemical systems being used *(1)*. One is a signal exchange system, such as a mixture of Concanavalin A and glycosylated insulin, or a mixture of boron- and diol-containing polymers. These chemical combinations utilize the specific interactions between lectin and glucose, or between boron and diol, respectively. The other system is a signal transduction which uses glucose-sensitive enzymes. The system transduces the glucose signal to other physicochemical signals, such as pH or redox. The enzymes are coupled to stimuli-responsive chemical systems.

We have designed and synthesized pH sensitive and redox-sensitive membrane systems. The former is a porous cellulose membrane on which poly(acrylic acid) and glucose oxidase (GOD) were immobilized *(2)*. When glucose is present, GOD catalyzes the conversion of glucose to gluconic acid, thus leading to a lower pH. The decrease in pH results in conformational modifications in the graft polymer chain which, in turn, alters the size of the membrane pores *(3,4)*.

The redox sensitive system is a composite membrane on which insulin is immobilized via a disulfide linkage. When glucose is oxidized by glucose dehydrogenase (GDH) in solution, the disulfide bonds are cleaved and the insulin is released *(5,6)*. Since GDH has no redox site, the coupling with FAD and NAD enhances the sensitivity.

In this paper a protein device was developed, in which GOD was directly coupled with insulin through a disulfide bond. The principal mechanism for the

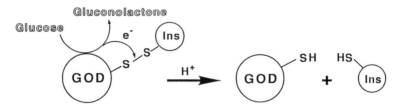

Figure 1. The principle of glucose-sensitive insulin-releasing protein device.

release of insulin is shown in Figure 1. When GOD was used, the enzyme had two active sites containing FAD. Therefore, an S-S bond was incorporated directly onto the protein. According to Degani and Heller *(7)*, although the active site is isolated from the protein surface, the electron transfer from the sites to the surface can be performed by coupling ferrocene derivatives in the presence of urea. In this study, the reversibly-cleavable S-S bonds are incorporated into GOD instead of the ferrocene derivatives.

Experimental

Materials. The glucose oxidase enzyme (GOD) from *Aspergillus niger* (G-8135) and insulin (I-5500) were purchased from Sigma Co. 5,5'-Dithiobis(2-nitrobenzoic acid) (DTNB) and 1-ethyl-3-(3-dimethylaminopropyl)carbodiimide hydrochloride, which is a water soluble carbodiimide (WSC), were purchased from Tokyo Kasei (Tokyo, Japan).

Preparation of Protein Devices. The protein device was synthesized according to the scheme shown in Figure 2. The insulin carboxylic groups were blocked by a methyl ester to prevent inter- and intramolecular cross-linking reactions of proteins in the WSC-activated reaction between insulin and DTNB, or the DTNB-insulin and GOD. The protected insulin was purified by GPC (Sephadex G-15). The DTNB was activated with WSC (10 wt%) and reacted with the insulin methyl ester at 4°C for 20 h. The reaction product was purified on the same GPC column. The insulin methyl ester modified with DTNB (DTNB-insulin) was reactivated with WSC (10 wt%) and reacted with GOD, which was pretreated with urea (2M) at 0°C for 2 h. The reaction product was purified by GPC to collect the insulin/GOD hybrid.

Insulin Release Assay. After the aqueous saccharide solution was added to the insulin-GOD hybrid solution, a portion of solution was removed for analysis. The amount of the insulin released was determined by the 276 nm absorption intensity of the solution eluted from a reverse-phase column packed with a Biofine RPC-PO (JASCO, Tokyo, Japan). A calibration curve based on the known amounts of native insulin was used in the insulin release assay.

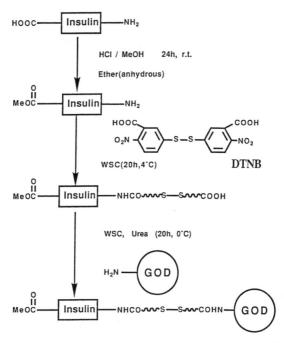

Figure 2. Scheme for the preparation of the protein device.

Biological Activity Assay of the Released Insulin. The biological activity of the released insulin was determined by measuring the amount of glucose taken up by the adipocytes. Albumin-containing Krebs-Ringer bicarbonate buffer solution (1 mL) was added to an adipocyte culture (mouse origin). The mixture was incubated for 2 h in the presence of glucose/^{14}C-glucose and insulin derivatives. The culture was terminated by adding sulfuric acid. Fat containing ^{14}C-glucose, which was taken up by the cells, was extracted with a toluene scintillation solution. To assess the relative biological activity of the insulin derivative, the amount of ^{14}C was determined by the scintillation counting and compared with that for the incubation with the native insulin.

Results

The elemental analysis of the insulin product after the DTNB reaction (molar ratio 1/5) yielded 3.36% sulfur, indicating that two or three amino groups of the insulin molecule were used in the reaction.

Summarized in Table 1 are the results for the reaction of DTNB-insulin and GOD. The increasing feed concentration of DTNB-insulin in the reaction with GOD resulted in the increase of DTNB-insulin incorporated into GOD.

Table 1. The Amount of Insulin Coupled with Glucose Oxidase

Molar ratio of DTNB-insulin/ GOD in feed*	Content of S in protein hybrid (%)	GOD amino groups used for DTNB-insulin coupling (mol%)
10 / 1	3.39	38
1 / 1	2.27	20
1 / 10	1.67	11

* DTNB-insulin was prepared by the reaction of DTNB and insulin in 5/1 molar ratio.

Insulin Release. When glucose was added to an aqueous solution containing the protein device, the HPLC pattern showed a new peak which was not distinguishable from native insulin, as shown in Figure 3. This result indicates that there was no disulfide bond cleavage in the insulin.

As shown in Figure 4, sugars other than glucose did not induce the insulin release from the insulin/GOD hybrid, indicating that the insulin/GOD hybrid has glucose specificity. On addition of a reducing agent, 2-mercaptoethanol, insulin was released in a manner similar to that for the glucose addition. This suggests that the release mechanism is based on the reduction cleavage of the disulfide bond between the insulin and the GOD.

Shown in Figure 5 is that insulin was repeatedly released on further addition of glucose. The protein hybrid with a higher amount of the incorporated insulin released more insulin for the same amount of the glucose added. The amount of the insulin released per glucose molecule ranged from 0.12 to 0.28 x 10^{-2}. The efficiency was higher for the insulin hybrid with the higher insulin content.

Biological Activity of the Released Insulin. The biological activity of insulin derivatives is demonstrated in Figure 6. The activity of the insulin methyl ester and DTNB-insulin are about 95% and 70% of the native insulin activity, respectively. The activity of the insulin released was about 80% of that for the native insulin. The HPLC pattern of the released insulin indicates no reduction cleavage of the disulfide bond in the insulin molecule.

Discussion

We have synthesized a prototype protein device which quickly releases insulin in response to glucose. However, the practical application of the device requires the

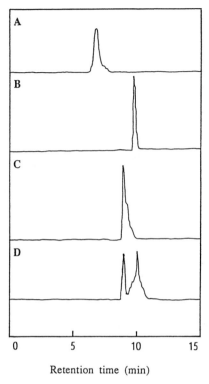

Figure 3. HPLC pattern of proteins. A - insulin; B - glucose oxidase; C - protein device; D -released insulin.

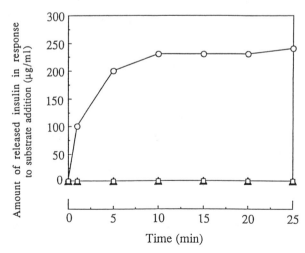

Figure 4. Insulin release from the protein device in the presence of 2-mercaptoethanol (▲), maltose (□), galactose (○). All concentrations are 15 mM.

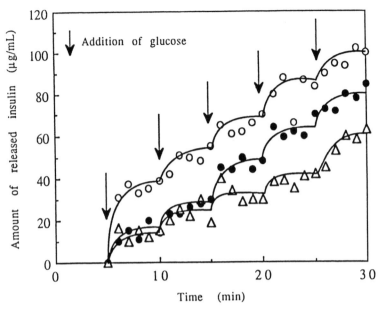

Figure 5. Insulin release from various protein devices in response to 15 mM glucose at 37°C. Protein hybrid synthesized with a molar ratio of DTNB-insulin/GOD in feed: 10/1 (O), 1/1 (●), 1/10 (△).

consideration of two other problems. The one is the immunogenicity of the device. The other is the lack of a need-response function for the device. The former can be overcome by encapsulation using semipermeable membranes, which are permeable to insulin but not to any immunoglobulin or covalent bonded bioinert polymers, such as poly(ethylene glycol). In the latter case, the device releases insulin in an almost linear manner in response to the glucose concentration, as shown in Figure 7. However, in the body insulin is only required at high levels of glucose in the blood, while at low glucose levels insulin release can cause serious problems. Normally, a healthy pancreas controls these conditions precisely.

In general, the intelligent materials used under similar biological conditions should have three functions: sensing, processing, and reacting. However, the processing function, which is usually performed by microcomputers in an electromechanical artificial pancreas, has not been included in the chemical systems. We, therefore, have introduced a processing control function into this system. To reduce the insulin release at the low glucose level in the blood, an inhibitor was incorporated in the protein device, as shown in Figure 7. We believe that this approach will result in more sophisticated artificial pancreas designs.

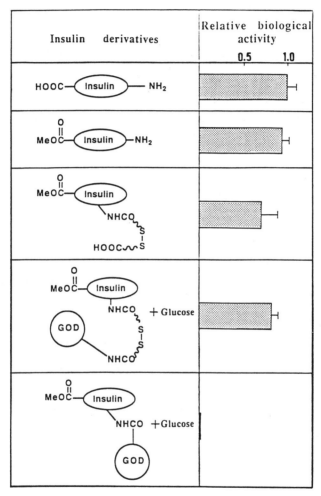

Figure 6. Biological activity of native, esterified, DTNB-modified, and released insulin from the protein hybrid in response to 15 mM glucose added. The activity of native insulin is taken as the standard.

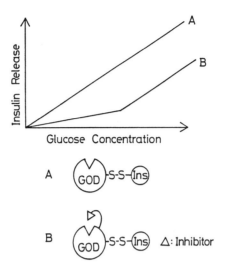

Figure 7. Scheme of the design for sophisticated glucose-sensitive insulin-releasing system.

Literature Cited

1. Ito, Y. In *Synthesis of Biocomposite Material: Chemical and Biological Modifications of Natural Polymers.* Imanishi, Y., Ed., CRC Press, Boca Raton, 1992, 137-180.
2. Ito, Y., Casolaro, M., Kono., K., Imanishi, Y. *J. Controlled Release* **1989**, 10, 195-203.
3. Ito, Y., Kotera, S., Inaba, M., Kono., K., Imanishi, Y. *Polymer* **1990**, 31, 2157-2161.
4. Ito, Y., Inaba, M., Chung, D.J., Imanishi, Y. *Macromolecules* **1992**, 25, 7313-7316.
5. Ito, Y., Chung, D.J., Imanishi, Y. *Artif. Organs* **1990**, 14, 234-236.
6. Chung, D.J., Ito, Y., Imanishi, Y. *J. Controlled Release* **1992**, 18, 45-54.
7. Degani, Y., Heller, A. *J. Am. Chem. Soc.* **1988**, 110, 2615-2620.

RECEIVED September 30, 1993

Chapter 5

Preparation and Characterization of Enzyme-Digestible Hydrogels from Natural Polymers by Gamma Irradiation

Kalpana R. Kamath and Kinam Park

School of Pharmacy, Purdue University, West Lafayette, IN 47907

Enzyme-digestible hydrogels were prepared from modified natural polymers without the addition of external crosslinking agents. Gelatin and dextran were first functionalized to introduce double bonds. The functionalized polymers were purified and then gamma-irradiated to form hydrogels. Swelling kinetics of these hydrogels was studied in the absence and presence of respective enzymes. The degradation of the hydrogels was prolonged as the gamma-irradiation dose and/or the polymer concentration increased. A model protein, invertase, was incorporated in these hydrogels by mixing it with the functionalized biopolymers before gamma-irradiation. The effect of gamma-irradiation on the enzyme was studied by determining the bioactivity of the invertase inside the gels. Preliminary data indicate that the gamma-irradiation dose used for gel formation did not affect the biological activity of invertase.

The use of biodegradable hydrogels in drug delivery has been explored extensively in the past several years. Innovative approaches have been used to develop new, simple methods for the preparation of hydrogels. Hydrogels made from natural polymers such as proteins or polysaccharides are particularly useful for preparing biodegradable drug delivery systems. Such systems are susceptible to enzymatic digestion in the body. Inert biopolymers are usually modified to introduce reactive groups which form crosslinks upon activation. Dextran *(1)*, starch *(2,3)*, albumin *(4,5)*, and gelatin *(6)* have been commonly used for preparing biodegradable hydrogels.

Incorporation of high molecular weight proteins or peptide drugs in the hydrogels is generally carried out either by swelling of dried hydrogels in the protein solution or by preparing the hydrogel in the presence of proteins. In the former method the extent of drug loading in the hydrogel may be limited due to the restricted penetration of the proteins into the hydrogel. In the latter method,

0097–6156/94/0545–0055$08.00/0

hydrogels may be formed by polymerization of monomers or by crosslinking water-soluble polymer molecules. The crosslinking of polymer molecules usually involves use of external crosslinkers which may have an adverse effect on the biological activity of the protein drug being incorporated.

The objective of this work was to prepare hydrogels from natural polymers such as gelatin or dextran without externally added crosslinking agent. Our hydrogels were designed to incorporate high molecular weight protein moieties in such a way that the biological activity of the incorporated protein did not change significantly. The biopolymers were first functionalized by introducing double bonds and then crosslinked using gamma-irradiation. A model protein, invertase, was incorporated in these hydrogels and the bioactivity of the loaded invertase was determined. The hydrogels formed in this manner do not require further purification since an external crosslinker is not added during gel formation.

Materials and Methods

Preparation of Functionalized Gelatin and Dextran. Natural polymers were functionalized by introducing double bonds following the procedure established in our laboratory *(7)*. Functionalization of other proteins *(8-10)* and polysaccharides *(1,11,12)* has been reported in the literature. Calfskin gelatin (Polyscience) was dissolved in phosphate-buffered saline solution (PBS, pH 7.2) to obtain the final concentration of 50 mg/mL. Glycidyl acrylate (Aldrich) was added directly to this solution while stirring. The reaction was allowed to proceed with constant stirring at room temperature. After 49 h, 20% (w/v) glycine solution was added to stop the functionalization reaction. This solution was then dialyzed against PBS to purify the functionalized gelatin from other additives.

The degree of gelatin functionalization was determined by measuring the free amine groups of gelatin using 2,4,6-trinitrobenzenesulphonic acid *(13)*. In our experimental conditions, 87% of the total amine groups in gelatin available for the titration was functionalized. The concentration of gelatin was determined spectrophotometrically. Absorptivity determined for 0.1% w/v gelatin solution at 280 nm was 0.112. Functionalization of dextrans (MW 480,000 and 2,000,000) (Sigma) was carried out in a similar manner. The content of acrylic groups in the functionalized dextran was determined spectrophotometrically by measuring the decrease in absorbance at 480 nm in methanol-water (80:20) solution of bromine (0.2% v/v) *(1,14)*. Acrylamide was used as a standard for this assay. The degree of functionalization was found to depend on the molecular weight of dextran. One acrylic group was introduced per 1.83 glucose residues in dextran of molecular weight of 480,000. The extent of functionalization in dextran of molecular weight of 2,000,000 was one acrylic group per 10.9 glucose residues.

Preparation of Hydrogels. The functionalized gelatin or dextran was gamma-irradiated to prepare hydrogels. Prior to irradiation, the samples were degassed and purged with nitrogen. The samples were irradiated for desired time periods at the dose rate of 0.0804 Mrad/h. The irradiation time used for preparation of gels from

30 mg/mL of functionalized gelatin was varied from 1 h to 16 h. To examine the effect of concentration on gel formation and swelling properties, 15 mg/mL of functionalized gelatin was also gamma-irradiated for 2 h, 4 h, and 6 h. Functionalized dextrans (50 mg/mL) prepared from molecular weights of 480,000 and 2,000,000 were irradiated for 4 h, 6 h, and 8 h to form gels. The gels were then washed in water for 3 h, air-dried for 24 h and oven-dried at 60°C for 12 h.

Swelling Studies. The effects of gamma-irradiation dose on hydrogel swelling and enzyme-induced digestion were studied by examining the swelling kinetics. The effect of pepsin on the degradation of gels formed by different gamma-irradiation doses was examined by swelling the gels in the simulated gastric fluid both in the absence and presence of pepsin (Sigma) at 37°C. The swelling kinetics of dextran gels was studied in the absence and presence of dextranase (Sigma) in pH 6 phosphate buffer. The swelling ratio, Q, was calculated from the following equation.

$$Q = W*/W$$

where **W*** and **W** are the weights of the swollen and dry gels, respectively *(15)*.

Incorporation of Invertase into Gelatin Gels : Invertase, the enzyme which breaks down sucrose into glucose and fructose, was incorporated in the functionalized gelatin gels. Invertase was mixed with functionalized gelatin (30 mg/mL) and the gels were formed by using the method described above. Figure 1 shows the schematic description of hydrogel formation in the presence of invertase. The

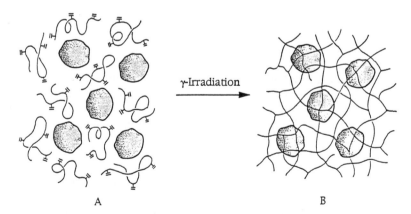

γ-Irradiation

A   B

Figure 1. Schematic description of invertase incorporation into hydrogels. Invertase (45 units/ml) was mixed with 30 mg/ml of functionalized gelatin (gelatin with double bonds) (A) and then gamma irradiated to form gelatin gels containing invertase (B).

activity of the incorporated enzyme was 45 units/mL of gelatin. The gamma-irradiation time used for the formation of gels was either 4 h or 6 h. The effect of gamma-irradiation on the bioactivity of invertase was determined by monitoring the amount of glucose generated in the presence of sucrose. Gels containing invertase were incubated in acetate buffer (pH 4.5) and 0.3 M sucrose at 55°C. After 1 h of incubation, the amount of glucose released by the action of invertase on sucrose was determined by using Glucose (HK) reagent (Sigma). The positive and negative controls used in this assay were the native invertase and the functionalized gelatin gels without invertase, respectively. The controls were treated in the same manner as the test samples.

Results and Discussion

The ability of the functionalized polymers to form hydrogels was dependent on the gamma-irradiation dose, concentration, and molecular weight of the polymer. For functionalized gelatin of 30 mg/mL concentration, the minimum irradiation time required for forming a network of desired integrity was 1 h. When the concentration was reduced to 15 mg/mL, hydrogel was formed after 2 h. The integrity of the gels increased as the gamma-irradiation dose increased. The formed gels did not dissolve even after incubation at 60°C for 12 h. This confirmed the crosslinking of gelatin molecules by gamma-irradiation. Functionalized dextrans (50 mg/mL) of 480,000 and 2,000,000 daltons formed gels after 4 h of gamma-irradiation. It appears that the rate of gel formation was dependent on the concentration of the biopolymers and the gamma-irradiation dose. As the concentration of polymers increased, gel was formed faster.

The swelling characteristics of these gels in the presence and absence of enzymes were affected by the irradiation time and the polymer concentration. Figure 2 shows the swelling kinetics of gelatin hydrogels formed from a concentration of 30 mg/mL. It was clear that the extent of gel swelling was dependent on the gamma-irradiation dose. As the irradiation time increased from 1 h to 16 h, the swelling ratio decreased. This is due to the higher crosslinking density that occurs at increased gamma-irradiation time. Little difference, however, was observed during swelling up to 40 h among gels formed by gamma-irradiation between 4 h and 16 h. For the loosely crosslinked gels which were prepared by gamma-irradiation for 1 h and 2 h, it was difficult to handle the swollen gels after about 60 h of incubation due to high water uptake.

Shown in Figure 3 are the swelling kinetics of the gelatin gels in the presence of pepsin at the activity of 250 units/mL. Gels obtained by gamma-irradiation for 1 h had a transient maximum swelling ratio at 3 h of incubation. The swelling ratio, which was about 12, was larger than that obtained in the absence of pepsin at the same time point. This indicates that bulk degradation occurs in the presence of pepsin. As the gamma-irradiation time increased, the time to reach the transient maximum increased. This confirmed that higher degrees of crosslinking at increased gamma- irradiation time make the gels more resistant to enzymatic attack.

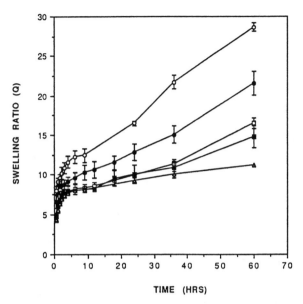

Figure 2. Swelling kinetics of gelatin hydrogels in the absence of pepsin. The hydrogels were prepared from 30 mg/ml of gelatin by gamma irradiation for 1 h (o), 2 h (●), 4 h (□), 6 h (■), and 16 h (Δ) (n=4). (Dose rate : 0.0804 Mrad/h).

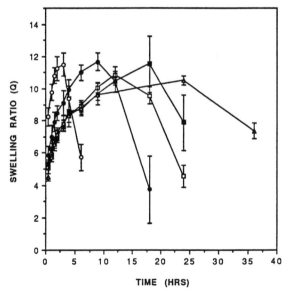

Figure 3. Swelling kinetics of gelatin hydrogels in the presence of pepsin. The hydrogels were prepared from 30 mg/ml of gelatin by gamma irradiation for 1 h (o), 2 h (●), 4 h (□), 6 h (■), and 16 h (Δ) (n=4).

The effect of gelatin concentration on the swelling kinetics in pepsin-free simulated gastric fluid is shown in Figure 4. As shown in the figure, concentration had a marked effect on the swelling properties of the gels. The swelling ratios of gels prepared from 15 mg/mL of functionalized gelatin were at least 1.5 times higher than those prepared from 30 mg/mL of gelatin. Figure 5 depicts the swelling of these two sets of gels in the presence of pepsin. The gels of 15 mg/mL gelatin showed a Q value of about 15 in the first 30 min, as compared to a Q value of 6 for gels of 30 mg/mL gelatin at the same time point. The transient maximum swelling ratios were much higher for 15 mg/mL of gelatin for all gamma-irradiation times. The effect of concentration was more obvious at lower gamma-irradiation times. The enzymatic degradation of the gels prepared from 15 mg/mL of gelatin occurred at a significantly faster rate than those prepared from 30 mg/mL of gelatin. The gels prepared from 15 mg/mL of gelatin by 2 h of gamma-irradiation degraded in about 3 h while those prepared from 30 mg/mL of gelatin took almost 18 h for complete degradation. This indicates that the lower concentration resulted in decreased crosslinking density which led to enhanced enzymatic digestion.

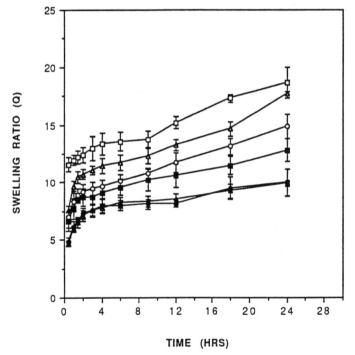

TIME (HRS)

Figure 4. Comparison of the swelling kinetics of gelatin hydrogels in the absence of pepsin. The hydrogels were prepared from 15 mg/ml (open symbols) or 30 mg/ml of gelatin (closed symbols) by gamma irradiation for 2 h ($\square$,$\blacksquare$), 4 h ($\triangle$, $\blacktriangle$), and 6 h (o, $\bullet$) (n=4).

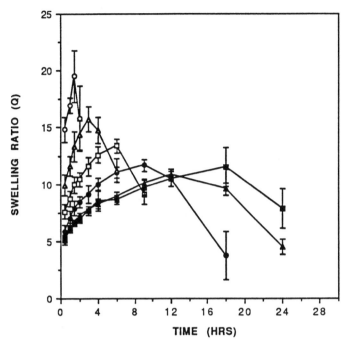

Figure 5. Comparison of the swelling kinetics of gelatin hydrogels in the presence of pepsin. The hydrogels were prepared from 15 mg/ml (open symbols) or 30 mg/ml of gelatin (closed symbols) by gamma irradiation for 2 h (o, ●), 4 h (△, ▲), and 6 h (□,■) (n=4).

Swelling studies of functionalized dextran gels also showed dependency of swelling kinetics on the gamma-irradiation time. However, the effect of gamma-irradiation time on the swelling properties was not as significant as that observed with the gelatin gels. The gels prepared from 480,000 dalton dextran showed lower swelling ratios as compared to those prepared from 2,000,000 dalton dextran (Figure 6). This particular swelling pattern may be attributed to the higher degree of functionalization of 480,000 dalton dextran than the 2,000,000 dalton dextran. Thus, the former sample forms more crosslinks during gamma-irradiation. This is confirmed by the dextranase-induced digestion studies for the functionalized dextran gels. Figure 7 shows the swelling kinetics of these two sets of gels in the presence of dextranase at 0.05 units/mL of activity. The higher molecular weight dextran gels showed a slightly higher transient maximum swelling than those prepared from 480,000 dalton dextran. The results of the swelling studies showed that factors such as the concentration and/or the molecular weight of the biopolymers could be adjusted to obtain systems with certain properties.

Shown in Figure 8 are the effects of gamma-irradiation on the ability of invertase to produce glucose as one of the products of sucrose breakdown. Sucrose

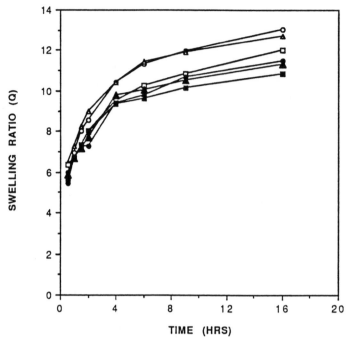

Figure 6. Comparison of the swelling kinetics of dextran hydrogels in the absence of dextranase. The hydrogels were prepared from dextran by gamma irradiation for 4 h (o, ●), 6 h (Δ, ▲), and 8 h (□,■). The concentration of dextran was 50 mg/ml and the molecular weights were 2,000,000 (open symbols) and 480,000 daltons (closed symbols) (n=4).

present in the surrounding medium entered the gel structure and the glucose produced by the action of invertase on sucrose diffused out of the gel. The amount of glucose released was determined as a measure of bioactivity of invertase. It should be noted, however, that this study was aimed at the determination of invertase activity incorporated in the gels and would not give information about the release of invertase from the gels. It was found that the bioactivity of invertase incorporated in the gelatin gels was not affected significantly by the gamma-irradiation for up to 6 h. Studies have also been conducted by other investigators *(16,17)* on the incorporation of invertase in hydrogels of synthetic polymers prepared by gamma- irradiation. A significant loss in invertase activity was reported due to high gamma- irradiation dose (4-7 Mrad) used for gel formation. Results of our studies indicate that proteins can be effectively incorporated in the hydrogels without compensating for their activity if the gamma-irradiation dose is kept low. In summary, hydrogel formation by gamma-irradiation of functionalized polymers is an effective way of incorporating bioactive proteins.

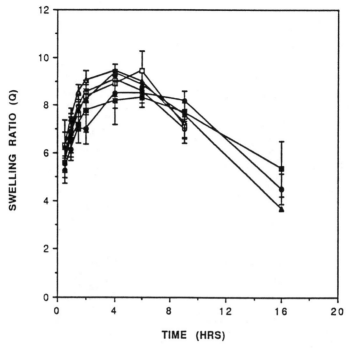

Figure 7. Comparison of the swelling kinetics of dextran hydrogels in the presence of dextranase. The hydrogels were prepared from dextran by gamma irradiation for 4 h (o, ●), 6 h (□, ■), and 8 h (△, ▲). The concentration of dextran was 50 mg/ml and the molecular weights were 2,000,000 (open symbols) and 480,000 daltons (closed symbols) (n=4).

Conclusion

Our approach of hydrogel preparation is a simple and efficient method which can be applied to a variety of water-soluble polymers. The functionalized water-soluble polymers can be purified before hydrogel formation. This obviates the need for purification of the formed hydrogels. In addition, use of an external crosslinking agent is not necessary for network formation. The method is rapid as compared to the chemical crosslinking method of gel preparation. The properties of the hydrogels are affected by parameters such as the gamma-irradiation dose, type of biopolymer, polymer concentration, and molecular weight of the biopolymer. This provides flexibility in the preparation of biodegradable hydrogels. One of the important advantages of this method is that the drug molecules could be mixed with functionalized water-soluble polymers before hydrogel formation by gamma-irradiation. The studies indicated that it was possible to incorporate proteins in these gels without adversely affecting their biological activity. This approach will be of particular interest in the loading of large molecular weight moieties such as peptides or protein drugs into the hydrogel drug delivery systems.

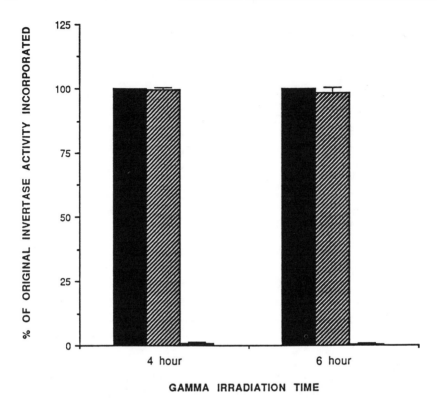

Figure 8. Effect of gamma irradiation on the bioactivity of invertase. The activity of the incorporated invertase was 45 units/ml. The concentration of gelatin was 30 mg/ml. The mixture was irradiated for 4 h or 6 h to form hydrogels. The bioactivity of invertase is expressed as percent of the original activity incorporated for native invertase (■), invertase incorporated in hydrogels (▨), and gelatin hydrogels without invertase (▤) (n=3).

Acknowledgment. This study was supported by the ICI Pharmaceuticals Group and in part by 3M basic research grant.

Literature Cited

1. Edman, P.; Ekman, B.; Sjöholm, I. *J. Pharm. Sci.* **1980**, 69, 838-842.
2. Heller, J.; Pangburn, S. H.; Roskos, K. V. *Biomaterials* **1990**, 11, 345-350.
3. Artursson, P.; Edman, P.; Laakso, T.; Sjöholm, I. *J. Pharm. Sci.* **1984**, 73, 1507-1513.
4. Lee, K. T.; Sokoloski, J. D.; Royer, G. P. *Science* **1981**, 213, 233-235.
5. Goosen, M. F. A.; Leung, Y. F.; Chou, S.; Sun, A. M. *Biomater. Med. Dev.; Artif. Org.* **1982**, 10, 205-218.

6. Tabata, Y.; Ikada, Y. *Pharm. Res.* **1989**, 6, 422-427.

7. Park, K. *Biomaterials* **1988**, 9, 435-441.

8. Torchilin, V. P.; Maksimenko, A. V.; Smirnov, V. N.; Berezin, I. V.; Klibanov, A. M.; Martinek, K. *Biochim. Biophys. Acta.* **1979**, 567, 1-11.

9. Plate, N. A.; Postnikov, V. A.; Lukin, N. Y.; Eismont, M. Y.; Grudkova, G. *Polymer Sci. U. S. S. R.* **1982**, 24, 2668-2671.

10. Plate, N. A.; Malykh, A. V.; Uzhinova, L. D.; Mozhayev, V. V. *Polymer Sci. U. S. S. R.* **1989**, 31, 216-219.

11. Guiseley, K. B. In *Industrial Polysaccharides: Genetic Engineering, Structure/property Relations and Applications,* Yalpani, M., Ed.; Elsevier Science Publishers B. V.: Amsterdam, 1987; pp 139-147.

12. Plate, N. A.; Malykh, A. V.; Uzhinova, A. D.; Panov, V. P.; Rozenfel'd, M. A. *Polymer Sci. U. S. S. R.* **1989**, 31, 220-226.

13. Snyder, S. L.; Sobocinski, P. Z. *Anal. Biochem.* **1975**, 64, 284-288.

14. Hoppe, H.; Koppe, J.; Winkler, F. *Plaste Kautsch.* **1977**, 24, 105.

15. Shalaby, W. S. W.; Park, K. *Pharm. Res.* **1990**, 7, 816-823.

16. Maeda, H.; Suzuki, H.; Yamauchi, A.; Sakimae, A. *Biotechnol. Bioeng.* **1974**, 16, 1517-1528.

17. Maeda, H.; Suzuki, H.; Yamauchi, A. *Biotechnol. Bioeng.* **1973**, 15, 607-610.

RECEIVED May 4, 1993

Chapter 6

Pharmacokinetics and Toxicity of a p-Boronophenylalanine–Cyclodextrin Formulation Delivered by Intravenous Infusion to Dogs

T. R. LaHann[1], W. F. Bauer[2], P. Gavin[3], and D. R. Lu[4]

[1]Center for Toxicology Research and College of Pharmacy, Idaho State University, Pocatello, ID 83209
[2]Idaho National Engineering Laboratory, Idaho Falls, ID 83415
[3]College of Veterinary Medicine, Washington State University, Pullman, WA 99164
[4]College of Pharmacy, University of Georgia, Athens, GA 30602

Boron-neutron capture therapy (BNCT) is a potentially important treatment for metastatic malignant melanoma. Effective BNCT requires the selective uptake of a boron compound by tumor cells. Para-boronophenylalanine (BPA) is of particular interest for BNCT since it is selectively delivered to melanoma tissue. However, its use is limited by its poor solubility in water at physiological pH. To facilitate the delivery of BPA to melanoma tissue sites, BPA's aqueous solubility was increased by forming a host-guest complex with 2-hydroxypropyl-β-cyclodextrin (HP-ß-CD). An *in vivo* study was carried out in dogs to examine BPA levels in plasma and selected tissues after i.v. infusion of BPA. Two BPA formulations were used: BPA in a pH 7.4 phosphate buffer and in an HP-ß-CD formulation. The pharmacokinetic profiles of BPA in both formulations were determined. The plasma concentrations of boron for the BPA/HP-ß-CD formulation were much higher than those for the BPA/buffer formulation. The area under each plasma boron concentration-time curve for BPA/HP-ß-CD was 20.7 times that for BPA/buffer. Thus, the delivery of BPA into the blood circulation was significantly enhanced due to the increase in BPA solubility.

The incidence of melanoma in Europe, Australia and North America has increased dramatically over the last twenty years. For example, in the U.S., the incidence of melanoma increased 83% between 1973 and 1987 *(1)*, and is currently increasing at a rate of about 3% per year *(2)*. Current therapies for malignant melanoma are effective in the early stages of the disease, but are of limited usefulness in later

0097–6156/94/0545–0066$08.00/0

stages *(3)*. Not surprisingly, new approaches to treating melanoma are of increasing interest. Recent animal and clinical experiments indicate that boron neutron capture therapy (BNCT) can successfully treat malignant melanoma *(4,5,6)* and may, in fact, be particularly effective in the treatment of large or metastasized tumors. The underlying concept of using BNCT for the treatment of cancer is that if a compound containing [10]boron can be selectively delivered to a tumor, the boron can be activated by an external neutron source. When the boron captures a slow moving (thermal) neutron, the absorbed energy causes the boron to fission into a lithium and a helium ion. These high energy fission products travel only very short distances, usually less than the diameter of a single cell. Mitochondria, lysosomes and DNA of the cells containing large numbers of boron atoms can be severely damaged by these fission products, and this, in turn, can lead to death of the cell.

Effective BNCT requires that 1) tumor cells selectively accumulate a [10]boron-containing drug, and 2) neutrons of sufficient velocity can penetrate to the level of the tumor to activate the boron. P-boronophenylalanine (BPA, Figure 1)

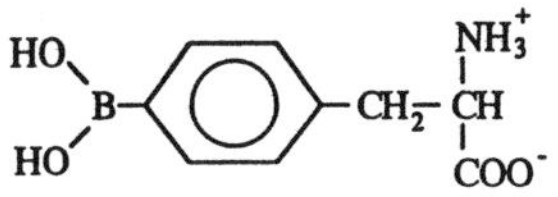

Figure 1. Chemical structure of p-boronophenylalanine.

is the drug most often used as a boron delivery system for malignant melanoma. BPA was initially proposed as a boron delivery drug because it was postulated that this amino acid analog, by mimicking a melanin precursor, would selectively accumulate in melanoma cells. Research indicates that BPA does selectively accumulate in melanocytes *(7,8)*, apparently being taken up by an amino acid transport system. Current evidence, however, suggests that BPA is not incorporated into melanin *(7,9)*. For successful BNCT, tumor boron concentrations of at least 20 ppm are thought to be necessary, but higher tumor boron levels are desirable. Calculations indicate that for a given neutron exposure, each doubling of the tumor boron concentration should increase tumor cell kill by a factor of about 10,000 *(10)*. Thus, even modest increases in the amount of boron in tumor cells can dramatically improve the effectiveness of BNCT as a cancer treatment. A major limitation of BPA is that, at physiological pH, it is poorly soluble in water, so that parenteral administration yields low and/or variable tumor boron levels. If uptake by an amino acid transport system is primarily responsible for accumulation of BPA, then high extracellular concentrations of BPA should increase the amount of BPA entering melanocytes. An i.v. infusion of BPA is the simplest route for delivering BPA to the tumor cells, but is not widely used because of BPA's poor solubility in aqueous solution (<2 ppm). To facilitate the delivery of BPA to tissue sites, we sought to increase BPA's aqueous solubility by forming a host-guest complex with 2-hydroxypropyl-β-cyclodextrin.

Cyclodextrins are produced by enzymatic degradation of starch and typically composed of 6, 7 or 8 glucose molecules linked by alpha-1,4 bonds *(11)*. Pharmaceutically useful cyclodextrins have a cone-like or a donut-like shape with a hydrophilic exterior and a hydrophobic interior. Because the interior of a cyclodextrin is a lipoidal microenvironment, small non-polar molecules that can fit within this interior cavity are readily solubilized. The solubility of large non-polar molecules can also be enhanced if the non-polar portion of the molecule fits within the lipophilic cavity of the cyclodextrin. Because no covalent bonds are formed, these "inclusion complexes" of cyclodextrin (host) and hydrophobic drug (guest) are readily reversible and in equilibrium with free drug in the aqueous environment. Molecusol®, an amorphous mixture of 2-hydroxypropyl-β-cyclodextrins, was selected as the host because published reports of acute and subchronic toxicity testing indicated that it was non-toxic even after repeated parenteral administration *(12,13,14).*

The ultimate goal of our work is to improve the selective delivery of boron to malignant melanoma tumor sites. The specific objective of the work reported here was to compare boron levels in blood after i.v. infusion of BPA formulated in 1) pH 7.4 phosphate buffer and 2) 45% 2-hydroxypropyl-β-cyclodextrin-water (w/v) and to determine the pharmacokinetic profiles of boron after i.v. infusion of these formulations to dogs. Unexpected adverse effects associated with intravenous infusion of 2-hydroxypropyl-β-cyclodextrin (Molecusol®) were observed and these effects are reported here.

Methods

Formulation of BPA. D,L-3-p-boronophenylalanine was purchased from Callery Chemical Company (Pittsburgh, PA). 2-Hydroxypropyl-β-cyclodextrin (Molecusol®) was purchased from Pharmatec, Inc. (Alachua, FL). A sample of 2-hydroxypropyl-γ-cyclodextrin (HP-γ-CD) was also provided by Pharmatec, Inc.   Both the Molecusol® and the HP-γ-CD were dissolved in distilled water. BPA solubility was determined in neutralized distilled water, 0.1 M phosphate buffer (pH 7.4), in 10, 20, 30 and 45% solutions of Molecusol®-water (w/v) and in a 50% solution of HP-γ-CD-water (w/v). Saturated solutions of BPA were prepared by adding excess BPA to each solution, and sonicating the resultant suspensions for 4-6 hours. The pH of some suspensions was transiently increased to about 10, while other suspensions were maintained near physiological pH. After sonication, all solutions were pH adjusted to 7.4 and filtered, first through Whatman #5 filter paper and then through 0.45 and 0.22 micron filters (Millipore). BPA content was calculated from boron levels determined by inductively coupled plasma atomic emission spectrometry.   All formulations were prepared within 3 days of use.   High performance liquid chromatography studies demonstrated that aqueous solutions of BPA were stable at physiological pH for at least 3 days and that the solubilization protocol did not cause decomposition of the BPA.

Pharmacokinetic and Toxicity Studies of BPA Formulations. Three healthy female Beagle dogs (8.3-11.0 Kg) and one Schnauzer dog (10.9 kg) afflicted with metastatic melanoma were used for the *in vivo* experiments. The dogs were initially anesthetized by i.v. injection of barbiturate, intubated and switched to isoflurane anesthetic.   An indwelling catheter was placed in the left saphenous vein for infusion of the BPA formulations, and the right cephalic vein was cannulated for blood sampling. Blood pressure was monitored by a cannula in the lingual artery, respiratory rate was monitored by a pressure transducer attached to the intubation cannula; rectal temperature was monitored by a YSI thermistor and maintained with a heating pad. Hematology and blood chemistry parameters were measured before and at selected time points following infusion of the test formulations. BPA formulations were infused into the Beagle dogs at a target rate of 17 mL/kg/h over a one h period. The Schnauzer received an infusion of BPA-Molecusol® at a target rate of 8.5 mL/kg/h. The lower infusion rate for the Schnauzer was a precautionary measure because the dog was in poor physical condition. Dogs remained anesthetized for the infusion period and for one hour post-infusion. Each Beagle dog received three treatments: 1) BPA in phosphate buffer, 2) BPA in Molecusol® and 3) Molecusol® alone. At least three weeks were allowed between the infusions to ensure clearance of boron. Blood samples were collected from the Beagle dogs immediately prior to, during (every 20 minutes) and post-infusion (target: 15 samples during a 72 h period). Duplicate plasma and serum samples were prepared from the whole blood samples. Urine was also collected at timed intervals through 72 h. Except for skin biopsies, tissue samples were not collected from these dogs. Tissue samples were collected from the Schnauzer six hours after a one hour infusion of BPA/Molecusol®. Blood and tissue samples were stored at -20° C until analyzed for boron.

Pharmacokinetic analysis of the data was conducted using a full-scale, compartment-type pharmacokinetic program (PharmK v.1.2.) running on a Mackintosh computer. Modeling of the data was based on the equation: $C = Q_1(e^{\lambda 1T}-1)e^{-\lambda 1t} + Q_2(e^{\lambda 2T}-1)e^{-\lambda 2t}$ for both the infusion and post-infusion periods, where Q_1 and Q_2 were the constant terms *(15)*. During the infusion period, $T = t$ (i.e. the time variable). Therefore, the equation became $C = Q_1(1-e^{-\lambda 1t}) + Q_2(1-e^{-\lambda 2t})$. During the post-infusion period, however, T became a constant (i.e. the duration of infusion). Assuming t' was the post-infusion time and thus equal to t minus T, the equation then became $C = R_1e^{-\lambda 1t'} + R_2e^{-\lambda 2t'}$, where $R_1 = Q_1(1-e^{-\lambda 1T})$ and $R_2 = Q_2(1-e^{-\lambda 2T})$. The data for the post-infusion period was initially used for model fitting based on the biexponential expression. The initial parameter estimation needed for the Chi-square regression fitting was obtained by the exponential stripping method. The resultant parameters ($R_1 = 1.58$, $R_2 = 0.98$, $\lambda 1 = 1.12$, $\lambda 2 = 0.29$ for the BPA/buffer formulation study and $R_1 = 12.18$, $R_2 = 3.79$, $\lambda 1 = 0.42$, $\lambda 2 = 0.07$ for the BPA/Molecusol® formulation study) were then used as the initial parameter estimation for data fitting of both infusion and post-infusion periods. The half-life, volume of distribution, area under the plasma concentration-time curve (AUC) and total body clearance were calculated. The values for AUC include both the infusion phase and the post-infusion phase. Unless otherwise specified, BPA and boron levels are reported as mean $\pm$ SEM. Evaluation of the statistical significance of differences in solution concentration and plasma boron concentration was determined by Student's T-test. The P values less than 0.05 were considered significant.

Results

BPA Solubility Studies. Levels of endogenous boron in distilled water and in 0.1 M phosphate buffer (pH 7.4) were below the limits of detection (equivalent to <0.008 mg/mL BPA). Depending on the lot of BPA used, the BPA solubility in pH 7.4 buffer ranged from 1.15 to 3.7 mg/mL. For a given lot, the aqueous solubility of BPA was the same for both pH 7.4 buffer and neutralized distilled water. The mean measured solubility of BPA in distilled water (all lots combined) was 2.5 $\pm$ 0.3 (N=4), while in pH 7.4 phosphate buffer, it was 2.2 $\pm$ 0.4 mg/mL (N=7). If the pH of the BPA-buffer suspension was briefly increased to about 10 and then returned to pH 7.4, more BPA was solubilized (6.0 $\pm$ 0.1 mg/mL, N=4). BPA solubility in Molecusol®-water solutions is shown in Figure 2. BPA solubility in aqueous solution increased with increasing concentrations of Molecusol® (p < 0.05) and since 45% was approximately iso-osmotic, this concentration was selected for *in vivo* experimentation. With the lot of BPA used, the BPA concentration in neutralized distilled water was 1.4 $\pm$ 0.1 mg/mL (N=3), while in the 45% Molecusol®, it was 6.4 $\pm$0.1 mg/mL (N=3). Transiently raising the pH of the BPA-Molecusol® suspension to about 10 and then readjusting the pH to 7.4 solubilized more BPA (11.5 $\pm$ 1.0 mg/mL, N=3). In pH 7.4 HP-γ-CD/water, 6.2 mg/mL BPA was solubilized. Neither the low concentration (1-2 mg/mL) BPA-buffer nor either of the BPA-Molecusol® formulations showed any signs of precipitation over a 7 day observation period. The high concentration (6 mg/mL) BPA-buffer formulation did show precipitation within a few hours of its preparation.

Pharmacokinetic Profile. The plasma boron concentrations, plotted as a function of time after i.v. infusion of BPA-buffer (BPA: 1.15 mg/mL) and BPA-Molecusol® (BPA: 5.7 mg/mL) formulations are shown in Figure 3. Two plots are displayed in the same figure for comparison. Target rates for infusion were 17 mL/kg/h and

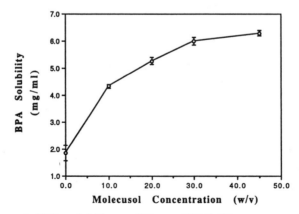

Figure 2. BPA solubility at different HP-ß-CD concentrations.

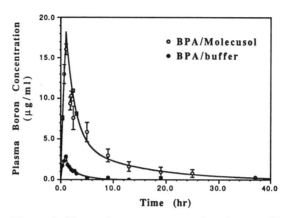

Figure 3. Plasma boron concentration-time profile.

actual rates of infusion were 17.4 $\pm$ 0.7 (BPA-buffer, N=3) and 16.2 $\pm$ 1.0 (BPA-Molecusol®, N=3) mL/kg/h. The mean infusion rate for the BPA-Molecusol® group was low because in one dog, unanticipated adverse reactions to the infusion necessitated termination of infusion after delivering only 82% of the intended dose. In earlier dog experiments, either serum or plasma boron levels were measured. In the experiments reported here, the boron content of both serum and plasma samples were measured. Significant differences between serum and plasma boron concentrations were not observed, indicating that either could be used in the study. The serum concentration of boron achieved at the termination of infusion of the BPA-Molecusol® formulation (16.5 $\pm$ 1.9 ppm B, N=3) was higher than that achieved with the BPA-buffer formulation (3.7 $\pm$ 0.2 ppm B, N=3) and higher than the peak measured values achieved with subcutaneous administration of a 200 mg/kg suspension (peak serum level: 3.4 $\pm$ 0.5 ppm B, N=5) or oral administration of a 50 mg/kg suspension (peak serum level: 2.0 $\pm$ 0.6 ppm B, N=3).

The data from both infusion and post-infusion periods were fitted to a two compartment, i.v. infusion model. The data is well described by the expression of $C = Q_1(e^{\lambda 1T}-1)e^{-\lambda 1t} + Q_2(e^{\lambda 2T}-1)e^{-\lambda 2t}$ (Figure 3), indicating that a two-compartmental i.v. infusion model does, in fact, constitute a suitable model. Pharmacokinetic parameters calculated from the model are listed in Table I.

The distribution of boron ($\lambda 1$-phase) delivered by the BPA-Molecusol® formulation ($t_{1/2}$ = 1.20 h) was slower than that for the BPA-buffer formulation ($t_{1/2}$ = 0.16 h). The initial fitting, based on post-infusion data only, showed that the distribution of boron ($\lambda 1$-phase) was similar for the two formulations ($t_{1/2}$ = 0.62 h for BPA/buffer formulation and $t_{1/2}$ = 1.64 h for BPA-Molecusol® formulation). Since only two data points were collected during the infusion period (0.33 and 0.67 h), this inconsistency requires further examination. Boron elimination was fairly rapid. The half-life in the $\lambda 2$-phase were 1.83 h for BPA-buffer formulation and

Table I. Pharmacokinetic Parameters Estimated from the Curve of Plasma Boron Concentration vs Time after i.v. Infusion[a,b]

Parameter	Formulation[c]	
	BPA/buffer (dose=0.96 mg B/kg)	BPA/Molecusol® (dose=4.43 mg B/kg)
Q_1 (μg/mL)	1.28	30.83
Q_2 (μg/mL)	5.67	59.27
R_1 (μg/mL)	1.27	13.50
R_2 (μg/mL)	1.79	4.67
λ_1 (h^{-1})	4.43	0.58
λ_2 (h^{-1})	0.38	0.08
$t_{1/2}$, λ_1 (h)	0.16	1.20
$t_{1/2}$, λ_2 (h)	1.83	8.43
Volume of distribution[d] (L/kg)	0.12	0.20
Area under curve[e] (μg h/mL)	6.95	90.10
Total body clearance[f] (L/h/kg)	0.14	0.05

[a]$C = Q_1(e^{\lambda 1T}-1)e^{-\lambda 1t} + Q_2(e^{\lambda 2T}-1)e^{-\lambda 2t}$
[b]$R_1 = Q_1(1-e^{-\lambda 1T})$ and $R_2 = Q_2(1-e^{-\lambda 2T})$
[c]B:BPA = 0.048
[d]Volume of distribution = Dose/$(Q_1\lambda_1 T + Q_2\lambda_2 T)$
[e]Area under curve was calculated based on the fitting results and included the infusion phase and the post-infusion phase.
[f]Total body clearance = dose/area under curve

8.43 h for BPA-Molecusol® formulation. These values were in good agreement with the values calculated from the post-infusion data ($t_{1/2}$ = 2.40 h for BPA-buffer formulation and $t_{1/2}$ = 9.90 h for BPA-Molecusol® formulation). When the dose of BPA was increased from 20.0 to 92.3 mg/kg, the half-life in the $\lambda 2$-phase increased about 4 times, suggesting that the pharmacokinetics of BPA-mediated boron are nonlinear in the dog model. In systems displaying nonlinear pharmaco-kinetics, total body clearance is expected to change as a function of dose and indeed, with increasing dose, the total body clearance of BPA-mediated boron decreased by 64%.

Tissue Levels of Boron and Urinary Excretion of Boron. The boron concentrations of tissues harvested 6 hours after i.v. infusion of BPA-Molecusol® to a dog afflicted with metastatic melanoma are shown in Table II. This animal received a BPA dose of 48.4 mg/kg over a one hour infusion period. The serum boron concentration at the termination of infusion was 5.6 ppm. While this was much lower than that levels achieved in non-tumor bearing beagle dogs (mean: 16.5 ppm B), the tumor-bearing Schnauzer received only half the amount of BPA administered to the Beagles. Uptake of BPA into tumor sites might also have contributed to the lower serum boron concentration in the Schnauzer. Tissue biopsies collected from Beagle dogs 1 hour after receiving BPA-buffer indicated skin/fat boron levels of <2 ppm. Analysis of cumulative boron excreted in the urine of the three dogs which received the BPA-Molecusol® formulation indicated that 60.8, 76.0 and 27.7% of the infused dose was excreted by 72 hours post-infusion. When the dogs were administered the BPA-buffer formulation, 72 hour excretion values were 57, 100 and 97% of the infused dose.

Table II. Tissue[1] Boron Concentrations 6 hours After I.V. Infusion of BPA-Molecusol®[2]

Tissue	Boron Concentration (ppm)
serum	1.2
plasma	1.1
pituitary	2.4
cerebral white matter	2.1
cerebral grey matter	1.9
cerebellum	2.4
thalamus	2.9
tongue	2.0
lung metastasis 1	4.9
lung metastasis 2	4.3
lung metastasis 3	18.8
lung (4 regions)	1.6-2.2
spleen	3.3
kidney metastasis	4.2
kidney	6.6
testicle	3.1

[1]Schnauzer dog afflicted with metastatic melanoma; Tissues with boron levels < 1.9 ppm: scalp, calvarium, medulla, cervical spinal cord, oral mucosa, thyroid, adrenal gland, temporalis muscle, forearm muscle, thigh muscle, stomach, duodenum, jejunum, cecum, colon and liver.

[2]Infusion conditions: 8.5 mL/kg/h, 1 h infusion, 5.7 mg BPA/mL.

Toxic Reactions Associated with BPA-buffer, BPA-Molecusol® and Molecusol® Only Infusions. The three dogs that received one hour infusions of BPA in buffer displayed no detectable adverse reactions, although there was a clinical impression that these dogs were somewhat slow in recovering from anesthesia. When the same three dogs received infusions of BPA-Molecusol®, more marked CNS depression was observed, i.e., the animals were very lethargic (not alert, displayed minimal locomotor activity) for up to 24 hours after termination of anesthetic. Infusion of the Molecusol® only formulation elicited much less lethargy than did the BPA-Molecusol® formulation, animals being up and around within one hour of termination of the gaseous anesthetic.

Infusion of the Molecusol® only formulation caused a transient (< 24 h) but marked (35-50%) reduction in white blood cell count and a rebound elevation (100-250% increase) of white blood cell count by 24 h post-infusion. Similar changes were seen after infusion of the BPA-Molecusol® formulation, but not after infusion of the BPA-buffer formulation. Infusion of BPA-buffer and BPA-Molecusol® formulations were associated with the release of nucleated red blood cells (nRBC) into the circulation. Relatively few of these immature red blood cells were noted after infusion of the BPA-buffer formulation (range: 0-2 nRBC/100 RBC), but after infusion of the BPA-Molecusol® infusion, as many as 41 nRBC were detected per 100 RBC's examined. Infusion of Molecusol® alone failed to stimulate detectable release of nRBC's into the circulation.

Dogs infused with BPA-buffer or Molecusol® only formulations displayed minimal changes in heart rate, rectal temperature and respiratory rate. Blood pressure did increase, but only during the course of the infusion, probably a result of the volume of fluid being infused. For example, all three dogs infused with the Molecusol® only formulation exhibited a moderate (20-40 mm Hg) increase in blood pressure which subsided to control values upon termination of the infusion. Three dogs were infused with the BPA-Molecusol formulation. One dog developed marked impairment of respiratory and cardiovascular function during the course of the infusion (Figure 4) and the infusion was terminated after 82% of the intended dose had been delivered. Infusion of the BPA-Molecusol® formulation into one dog was uneventful, while infusion into the third dog elicited a brief period of minor respiratory abnormalities reminiscent of those observed in the first dog, but not as severe. These alterations in respiration and cardiovascular function were not observed when the dogs were infused with Molecusol® alone. However, 2-3 days after the Molecusol® only infusion, moderately severe pulmonary edema was evident in two dogs. Veterinary care was unable to alleviate the pulmonary edema and the two dogs were euthanized.

Discussion

Our data demonstrate that for a given volume infused, formulation of BPA with a cyclodextrin delivers more boron into blood than does a formulation of BPA dissolved in pH 7.4 buffer. The ability of BPA to form a host-guest complex with Molecusol® depends both upon its lipophilicity and its ability to fit within the cyclodextrin's cavity. Molecusol® has a cavity diameter of 6-7.8 angstroms, while

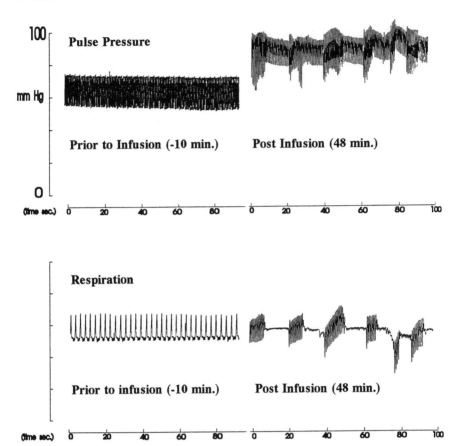

Figure 4. Cardiovascular and pulmonary function before and after i.v. infusion of BPA-Molecusol®. Infusion parameters: 17 mL/kg/h, 5.7 mg BPA/mL. Labeling: "prior to infusion" indicates arterial pressure and respiratory frequency 10 min before initiating the BPA-Molecusol® infusion; "post infusion" indicates arterial pressure and respiratory frequency 48 min into the infusion. Peaks on the respiratory tracing correspond to breaths; the top of the pulse pressure tracing represents systolic blood pressure, while the bottom of this tracing represents diastolic blood pressure.

HP-γ-CD has a cavity diameter of 7-9.5 angstroms. To determine if cavity size was limiting the cyclodextrin-mediated solubilization of BPA, we evaluated BPA solubility in HP-γ-CD-water. A 50% solution of HP-γ-CD failed to solubilize more BPA than did a 45% solution of Molecusol® (6.2 vs 6.4 mg/mL), so it appears that cavity size was not the limiting factor in solubilizing BPA.

The dynamic equilibrium that exists between free guest (BPA) and the host-guest complex should ensure the rapid release of free BPA as the BPA-Molecusol®

complex is diluted in blood. Molecusol®-mediated increases in the blood boron concentration, together with the presumed release of free BPA from the host-guest complex suggest that a BPA-Molecusol® formulation should enhance delivery of free BPA to tumor. BPA accumulation in melanocytes apparently occurs as the result of uptake via an amino acid transport mechanism, so delivery of more BPA to tumor sites should increase the accumulation of boron in malignant cells. However, data demonstrating that a BPA-cyclodextrin formulation actually enhances the tumor uptake of boron is not yet available.

The plasma boron concentration achieved with the BPA-Molecusol® formulation was much higher than the concentration achieved with the BPA-buffer formulation (p < 0.05). It was also higher than the levels achieved with subcutaneous or oral administration of BPA suspensions. Including both infusion and post-infusion times, the AUC from the BPA-Molecusol® infusion was 90.10 μg boron h/mL (N=3), while the AUC from the BPA-buffer infusion was 6.95 μg boron h/mL (N=3). Thus, by complexing BPA with Molecusol®, the delivery of BPA-mediated boron into the blood stream was increased almost 15-fold over that achievable with the BPA-buffer formulation. Analysis of the boron pharmacokinetic profile also indicated a nonlinear pharmacokinetic response in dogs. The reasons for this non-linearity are not yet clear and further studies are needed to evaluate the dose-dependency of the boron pharmacokinetics and to determine the extent to which Molecusol® may be contributing to this non-linearity. The plasma boron level achieved in the one animal afflicted with malignant melanoma was lower than that achieved in the non-tumor bearing dogs, but much of this difference reflects the lower dose of BPA administered. However, even with the infusion of a relatively small BPA dose (<50 mg/kg), boron concentrations in some metastatic tumor sites were high enough that boron neutron capture therapy could have been effective, at least at those localized sites.

Although formulation with a cyclodextrin increases the amount of BPA that can be delivered by i.v. infusion, the evidence also suggests that this formulation will elicit a higher incidence of systemic toxicities. Molecusol® was chosen as a host chemical because published reports indicate that it is a safe, clinically acceptable solubilizing agent. Toxicity studies conforming to the U.S. Good Laboratory Practices regulations have been conducted in rats, dogs and monkeys. In monkeys, acute i.v. injection of 10 g/kg of Molecusol® elicited no obvious toxic effects (14). Rats and dogs also tolerate large doses of Molecusol® well, although in rats, Molecusol® doses greater than 400 mg/kg have been associated with reversible vacuolization of the glomerular cells of the kidney (16). The side effects associated with infusion of the BPA-Molecusol® and Molecusol®-only formulations were unexpected. The extended recovery time in dogs dosed with BPA suggests that BPA may be acting as a central nervous system (CNS) depressant. CNS depression is not unique to the boron delivery drug BPA, for we have observed a similar effect with another boron delivery drug, borocaptate sodium.

The BPA-induced release of nucleated red blood cells into the circulation is interpreted as a non-specific response to stress. Because of the severity, the cardiopulmonary reactions associated with the BPA-Molecusol® formulation were particularly disturbing. To determine if the observed effects reflected a BPA-induced toxicity or a Molecusol®-induced toxicity, the dogs responding adversely to the BPA-Molecusol® formulation were challenged with a Molecusol®-only formulation. No cardiopulmonary abnormalities were observed, suggesting either that BPA was the causative agent or that both BPA and Molecusol® elicit sub-clinical but additive toxicities. Formulation of BPA as a BPA-fructose complex is also an effective i.v. delivery system for increasing tissue boron concentrations *(17)*. In general, rats tolerate i.v. infusions of BPA-fructose well. However, studies have revealed that a small percentage of rats respond to i.v. infusion of BPA by developing cardiovascular lability *(18)*. Failure to observe cardiopulmonary lability in dogs infused with Molecusol® only, plus the occasional observations of BPA-induced cardiovascular lability in rats, argue that the cardiopulmonary abnormalities observed in dogs may be at least partially attributed to BPA. However, rat studies also indicated that expression of BPA cardiovascular toxicities was stress-dependent *(18)*, and the results of the Molecusol® only infusion do suggest that the infusion of Molecusol® was a stressor.

Pulmonary edema was observed in two of three dogs infused with Molecusol®-only. There are no published reports of Molecusol®-induced pulmonary edema and pulmonary edema either did not occur, or was not of a sufficient magnitude to be detected in dogs treated with the BPA-Molecusol® formulation. Admittedly, dogs did receive a large amount of Molecusol® (7.9 g/kg over a 1 hour period). However, no clinically significant adverse effects were observed after i.v. injection of 10 g/kg to monkeys *(14)*, 5 g/kg to rabbits *(19)*, or rats (19) or 1 g/kg to dogs *(19)*. Other than suggesting that Molecusol® might have reduced the concentration of a critical endogenous substance (by formation of host-guest complexes), we can offer no explanation for the Molecusol®-induced pulmonary edema.

In summary, the results of this study indicate that i.v. infusion of an aqueous BPA-Molecusol® complex delivers higher plasma boron levels than can be achieved by infusion of BPA dissolved in pH 7.4 buffer. These results lead us to hypothesize that host-guest formulation of BPA will increase the delivery of boron to tissue (tumor) sites. However, studies also suggest that the potential for serious adverse reactions may limit the utility of the BPA-Molecusol® formulation.

Acknowledgements

The authors wish to thank Dr. E. Jarvi (ISU) for performing the HPLC analysis, and Dr. S. Kraft, Ms. C. Johnson and Mr. V. Sweet (WSU) and Ms. G. Daniell (ISU) for assisting with the animal experiments. This study was performed under the auspices of the U.S. Department of Energy, Office of Energy Research, DOE Idaho Field Office, Contract No. DE-AC07-76ID01570.

Literature Cited

1. Henderson, B., Ross, R., Pike, M. *Science* **1991**, 254, 1131-1138.
2. Davis, D., Hoel, D., Fox, J., Lopex, A. *Lancet* **1990**, 336, 474-481.
3. Kon, H. *New Engl. J. Med.* **1991**, 325, 171-182.
4. Mishima, Y., Ichihashi, M., Hatta, S., Honda, C., Yamamura, K., Nakagawa, T. *Pig. Cell Res.* **1989**, 2, 226-234.
5. Barth, R., Soloway, A., Fairchild, R. *Scientific American* **1990**, 263, 100-107.
6. Coderre, J., Kalef-Ezra,J., Fairchild, R., Micca, P., Reinstein, L., Glass, J. *Cancer Res.* **1980**, 48, 6313-6316.
7. Coderre, J., Glass, J., Fairchild, R., Roy, U., Cohen, S., Faud, I. *Cancer Res.* **1987**, 47, 6377-6383.
8. Hatta, S., Tsuji, M., Honda, C., Ichihashi, M., Mishima, Y. *In Neutron Capture Therapy*, Hanatanake, H., Ed. Nishimura Co., Nigata Japan, 1986, 303-305.
9. Zha, X., Bennett, B., Ausserer, W., Morrison, G. In *Progress in Neutron Capture Therapy for Cancer*, Allen, B., Moore, D., Harrington, B., Eds., Plenum Press, New York, 1992, 331-333.
10. Wheeler, F., Griebenow, M., Wessol, D., Nigg, D., Anderl, R. In *Basic Life Sciences: Clinical Aspects of Neutron Capture Therapy*; Fairchild, R., Bond, V., Woodhead, A., Eds., Plenum Press, New York, 1989, 50, 165-178.
11. Brewster, M., Simplins, J., Hora, M., Stern, W., Bodor, N. *J. Parent. Sci. Technol.* **1989**, 43, 231.
12. Seller, K., Szathmary, S., Huss, J., DeCoster, R., Junge, W. In *Minutes of the Fifth International Symposium on Cyclodextrins*, Dominiques Du Chenn, Ed., Editions De Sante, 1990, 518-521.
13. Coussement, W., Van Cauteren, H., Vandenberge, J., Vanparys, P., Teuns, G., Lampo, A., Marsboom, R. In *Minutes of the Fifth International Symposium on Cyclodextrins*, Dominiques Du Chenn, Ed., Editions De Sante, 1990, 522-524.
14. Brewster, M., Bodrol, N. In *Minutes of the Fifth International Symposium on Cyclodextrins*, Dominiques Du Chenn, Ed., Editions De Sante, 1990, 525-534.
15. Gibaldi, M., Perrier, D. In *Pharmacokinetics*, Marcel Dekker, New York, 1982, 45-111.
16. Personal Communication from R. Stratton, Pharmatec, Inc.
17. LaHann, T., Lu, D., Daniell, G., Sills, C., Craft, S,. Gavin, P., Bauer, W. In *Proceedings of the Fifth International Symposium on Neutron Capture Therapy*, Plenum Press, In Press.
18. LaHann, T., Sills, C., Hematillake, G., DymockT., Daniell, G. In *Proceedings of the Fifth International Symposium on Neutron Capture Therapy*, Plenum Press, In Press.
19. PHARMATEC Internal Records. Referenced in: Molecusol® HPB Safety Testing, Pharmatec, Inc. 1988, revised 6/28/90.

RECEIVED August 5, 1993

Chapter 7

Electrorelease of Drugs from Composite Polymer Films

Maria Hepel and Zbigniew Fijalek

Department of Chemistry, Potsdam College of the State University of New York, Potsdam, NY 13676

New composite polypyrrole films with a permanently incorporated biopolymer, Melanin, have been prepared. Melanin was incorporated into polypyrrole by a electropolymerization process. These materials have potential applications for electrochemically controlled drug release to the CNS or for transdermal drug delivery. The full characterization of the composite polymer films was obtained by *in situ* monitoring of mass changes using the quartz crystal microbalance in conjunction with cyclic voltametry. The EQCM allowed direct measurements of the amount of the drug released when the film potential was changed. It is possible to incorporate and subsequently release a variety of neuroleptic drugs such as phenothiazine derivatives and tricyclic antidepressants upon application of electrochemical pulse.

In recent years, a considerable amount of research has been carried out to evaluate alternative modes of drug administration. Much of the effort has been directed towards the development of controlled drug delivery systems *(1-10)*. Controlled release polymers or pumps offer a number of potential advantages when compared to conventional methods of drug administration, *e.g.,* injection, oral ingestion, etc. The main objective in the drug delivery systems is to achieve an effective therapeutic administration for an extended period of time and to provide the drug only when it is needed at the appropriate concentration for the desired therapeutic effect. The essential ingredients of such a system are a sensing mechanism that can detect concentrations in the blood and transfer this information to a delivery device that can then modify therapeutic drug concentrations.

A number of rate-controlled pharmaceuticals have been registered for human use such as: pilocarpine *(12)*, progesterone *(1)*, theophylline *(13)*, indomethacin *(14)*, scopolamine *(17)*, nitroglycerine *(16)*, and clonidine *(15)*. Examples of various techniques to control drug release are the following: pumps providing different flow rates, polymers responding to pH stimuli, polymers where

0097–6156/94/0545–0079$08.00/0

release of drug is activated by chemical reaction, by solvent or enzyme, polymers where release of drug occurs by diffusion, polymers that release the drug in response to temperature, an oscillating external magnetic field or ultrasound and polymers that release drug in response to electrical stimulation (18-28,37,40).

Advantages of controlled drug delivery systems are numerous. They include: better control of drug level resulting in fewer side effects, the opportunity to deliver the drug locally, decreased requirements for the total amount of drug, and protection of drugs that are rapidly destroyed by the body.

Drug systems that deliver an active agent to a specific body site in precisely regulated amounts are superior to those that indiscriminately flood the whole body with a therapeutic agent. Drug targeting is an area of great research interest (29-31). With implants, the localized action can be obtained by placing the device in the vicinity of the receptor organ and directing the drug selectively to the target tissue while preventing it from interacting with other tissues. Selective drug delivery to the CNS is of great importance (32-36), e.g., to deliver dopamine to the appropriate area of the brain to control Parkinson's disease. Iontophoresis is another method of choice. Iontophoresis is a method of transdermal drug delivery in which ionized drug compounds are driven into the body via electric current. Conductive polymers are attractive electrode materials for transdermal drug delivery applications because they are capable of transporting ionic drugs during redox processes (47). In our research, we have applied the electrochemical method, based on the piezoelectric sensor, for controlled drug release (37-38).

Controlled Drug Release by Electrochemical Pulse Stimulation

Electrochemical pulse stimulation for controlled drug release studies has many potential applications. As a matrix, conductive polymers that release drugs in response to electrical stimulation (37,39,40) have been employed in the controlled drug release systems. The advantage of using an electroactive polymer electrode lies in the fact that ionic drugs can be electrochemically loaded into and released from the solid polymeric host in a controlled process. The ability of the conductive polymer to bind ions is changed during its redox process. The change in the net charge on the redox polymer during reaction requires ions to flow into or out of the film and this allows the polymer film to bind and expel ions in response to the electrical signal. An application of particular interest is the drug delivery system where control of the potential, current or charge would allow the control of the amount of ionic drug that is released from a polymer. One should be able to bind anionic drugs into a cationic polymer, e.g., polypyrrole, and next force them out by neutralizing the charge on the polymer. When composite polymer films are used, then one should be able to bind cationic drugs. The polypyrrole films obtained by electropolymerization of pyrrole from solutions of simple inorganic electrolytes show predominant anionic dynamics. In order to control the uptake and release of cationic drugs, it is necessary to modify some properties of polypyrrole films, especially the ion dynamics. This goal can be achieved by doping polypyrrole with various ionic materials. Such a composite polymer material has cation-exchange properties and can act as a binder and releaser of cationic drugs.

Redox Switching Process

Reduction: $PPy^+(MeI^-) + e^- + CPZ^+ \longrightarrow PPy^o(MeI^-)(CPZ^+)$

Oxidation: $PPy^o(MeI^-)(CPZ^+) - e^- \longrightarrow PPy^+(MeI^-) + CPZ^+$

In this polymer composite which contains permanently incorporated polyanions, the electric charge compensation for oxidation and reduction of the polypyrrole backbone is achieved by the transfer of cations. Miller *et al.* *(40)* have prepared polymer composite, poly(N-methylpyrrolylium) poly(styrenesulfonate), which has the ability to bind and release dopamine. UV-visible spectroscopy was used to detect the amount of released drug.

In this paper we report the development of a new conducting polymer composite, polypyrrole/melanin, which can act as a cation exchange membrane, and the application of a new method for the drug release evaluation, *i.e.*, the EQCM technique

Experimental.

The electrochemical quartz crystal microbalance (EQCM) can be used as a sensitive piezoelectric sensor for the controlled drug release studies *(37)*. It allows for the simultaneous *in situ* determination of both the electrochemical parameters of the system and the effective mass of the electrode which makes the technique so attractive. AT-cut 10 MHz quartz crystals with vacuum evaporated gold were used as substrates for polymer electrodeposition. A Hewlett-Packard potentiostat and a plotter Model HP 7090A were used to control and record current-potential characteristics. An oscillator Model EQCN-500 (ELCHEMA, Potsdam, NY) was used to track resonant frequency of EQCM during the electrochemically induced mass changes. The relationship between frequency and mass for the simple case of a thin, rigid film is given by the Sauerbray equation:

$$\Delta f = - \frac{2 \Delta m n f_o^2}{A \sqrt{\mu_q d_q}}$$

In this formula, the change in the oscillation frequency (Δf) is equal to minus the change in interfacial mass (Δm) per unit area (A) times a constant. Thus, frequency decreases as the mass increases. The constant is evaluated with knowledge of the oscillation frequency of the fundamental mode of the EQCM (f_o), the overtone number (n), the density of quartz (d_q = 2.648 g cm^{-3}), and the shear modulus of quartz (μ_q = 2.947 x 10^{11} g cm^{-1} s^{-2}).

With careful circuit design and signal averaging, noise levels for the EQCM are typically below 1 Hz *(41)*. It is, therefore, possible to measure very small mass changes in the low nanogram range *(41)*. The EQCM is well suited for measurements of mass transport that can accompany redox processes that are occurring in thin films on electrodes *(41-45)*. Mass changes, associated with the gain or loss of any species in the film, were detected with the EQCM.

The linear relationship of the frequency change Δf to the mass change Δm for thin, rigid films makes the method simple and attractive. The EQCM can be used to monitor the movement of ionic species driven by the requirement to maintain electroneutrality in the bulk of the polymer film.

Results and Discussion.

Our research effort was focused on the preparation of novel, composite polymer materials. We studied polypyrrole composite films obtained by electropolymerization of pyrrole in the presence of suitable dopants. In the present study, we developed a new conductive polymer, which would act as a binder and releaser of cationic drugs. New polymer composite films, polypyrrole/melanin (PPy^+/Mel^-), with melanin permanently incorporated into the polypyrrole matrix during the polymerization process, were prepared. The stability of these films were tested using the EQCM technique. No leaching out of melanin incorporated during the polymerization process of pyrrole and no deterioration of such composite films were observed even after prolonged potential cycling.

The polymerization process was carried out at constant potential E = -600 mV vs. SCE from a solution containing 2×10^{-3} M pyrrole, 0.064 g/L melanin and 0.1 M NaCl. Significant differences in ion dynamics for polypyrrole/melanin composite in comparison to undoped polypyrrole films were observed. Melanin is a polyfunctional biopolymer consisting of indole-5,6-quinone units with several types of bonds (peroxide, ether, carbon-nitrogen, etc.) and with branched chains which form three-dimensional structures.

MELANIN

The ion dynamics of PPy/Mel films depends strongly on the film preparation conditions and is sensitive to the specific interactions of melanin with various components of the medium. It was of special interest to examine the degree of reversibility of these interactions. Therefore, a series of experiments was performed in which the dynamic characteristics of a single PPy/Mel film were determined in different electrolyte solutions by changing the medium and then washing the film and returning to the previous medium. These EQCM experiments show that the PPy/Mel films behave reversibly and after several potential cycles, the ion dynamics returns to that characteristic for the given electrolyte. A well defined cationic dynamics was observed for phosphate, sulfate and chloride environments, while in the nitrate medium, anionic dynamics becomes evident in more anodic potentials. The cationic dynamics observed for PPy/MeI films in phosphate, sulfate and chloride solutions changes reversibly to mixed anionic/cationic dynamics when medium is changed to nitrate solutions. When the medium is changed back to phosphate, sulfate or chloride solution, the cationic dynamics is quickly restored *(38)*.

The polypyrrole/melanin films have cationic dynamics and were capable of binding cationic drugs like chlorpromazine (CPZ) at potentials more negative than -200 mV vs SCE and release them when the potential was scanned in the positive direction.

$$CH_2CH_2CH_2N^+(CH_3)_2$$
$$H \qquad Cl^-$$

Chlorpromazine HCl

The shapes of the current-potential curves in the presence of 10^{-2} M CPZ in solution are different from those observed in 0.1 M NaCl. Also frequency-potential curves show larger hysteresis for CPZ solutions, indicating fast uptake of the drug at negative potentials and a slow release during the positive scan. (Compare curves in Figure 1 in the presence of 10^{-2}M CPZ with curves in Figure 2 for the 0.1 M NaCl solution without CPZ). This hysteresis can be explained by the larger sized CPZ cation compared to sodium and by stronger interactions of the former with melanin.

The presence of another complexing agent, EDTA, in solution, inhibits uptake and release of CPZ. This inhibitory effect can be explained by the formation of complexes between CPZ and EDTA. Competitive effects by other cations like Na^+, K^+, Ca^{+2}, Zn^{+2}, Cu^{+2} on the binding of CPZ was also studied. In most cases, simultaneous binding of interfering cation and CPZ was observed.

The CPZ uptake at -800 mV and the release at +200 mV upon the potential step application program is presented in Fig. 3. The data from this figure confirm a fast uptake of CPZ and a slow release. The effect of the height of the potential step causing the CPZ release and the presence of other drugs and metal ions on the release rate of CPZ was tested. Additional evidence for the release of CPZ from polymer film, previously loaded with CPZ in a solution containing only 0.1 M NaCl, was obtained from UV-Visible Spectroscopy data. Absorbance of solutions containing CPZ was measured at 252 nm. The mass of released CPZ calculated from UV-Visible Spectroscopy data was about 20% lower than the mass calculated form the EQCM experiments. The higher mass calculated from the EQCM data is explained by the combined effect of drug and solvent release.

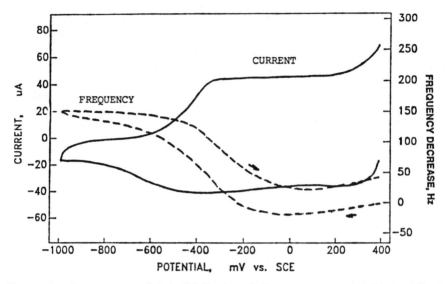

Figure 1. Current-potential (solid line) and frequency-potential (dashed line) characteristics recorded in 0.1 M NaCl solution with the EQCM electrode covered by electrodeposited composite film PPy/Melanin. Scan rate: 50 mV/s.

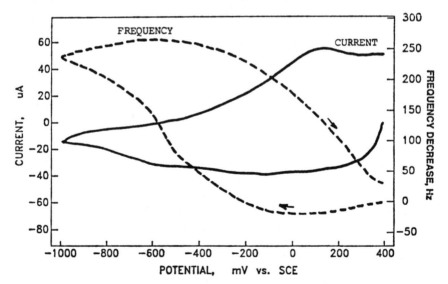

Figure 2. Current-potential (solid line) and frequency-potential (dashed line) characteristics recorded in 1×10^{-2} M chlorpromazine solution with the EQCM electrode covered by electrodeposited composite film PPy/Melanin. Scan rate: 50 mV/s; 25th scan.

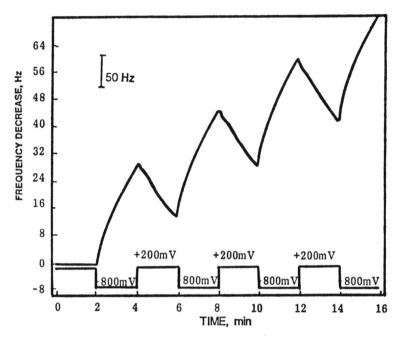

Figure 3. Frequency changes vs. time recorded upon the subsequent potential steps from -800 mV to +200 mV vs. SCE in solution containing 10^{-2} M chlorpromazine at the EQCM electrode covered with electrodeposited PPy/Melanin composite film.

The effects of the potential switching on the frequency change (mass change) for NaCl solution and CPZ solution are compared in Figures 4a and 4b. The potential was changed from E = +400 to E = -1000 mV vs. SCE and back to E = +400 with a pulse duration of t = 2 minutes. In NaCl solutions, the electrode mass increased (frequency decreases) virtually instantaneously upon the cathodic potential step (form E = +400 to E = -1000 mV) and mass stabilized quickly, while the mass changes observed for CPZ solutions were more complex. After a sharp mass increase, there was a slow steady increase in mass indicating that the uptake of CPZ cations was not complete during this short duration.

In Figure 5a the frequency change-time transient obtained for an EQCM PPy/Mel film in 0.01 M chlorpromazine solution, for a potential step from E = +200 mV to E = -800 mV vs. SCE, is presented. The sharp frequency decrease corresponding to the mass increase was observed. The electrode loaded with chlorpromazine was withdrawn from solution, washed and placed in 0.1 M NaCl solution. Subsequent release of the chlorpromazine from the PPy/Mel film was then determined. The frequency change-time transient shown in Figure 5b illustrates the dynamics of the chlorpromazine release which was forced by a potential step from E = -800 mV.

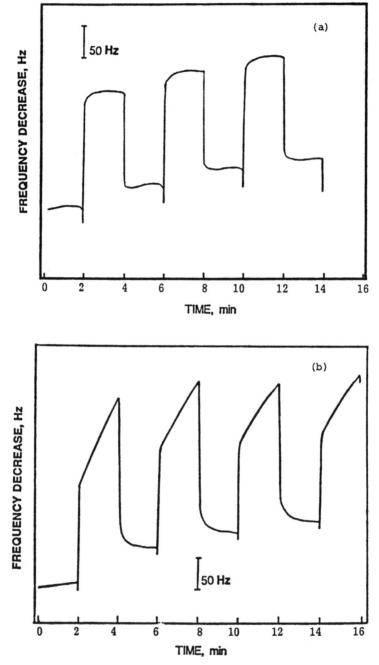

Figure 4. The frequency change-time transients obtained at an EQCM PPy/ Melanin modified gold electrode for potential steps every 2 minutes from E = 400 to E = -1000 mV vs. SCE in (a) 0.1 NaCl (pH = 6) solution, (b) 0.01 M CPZ.

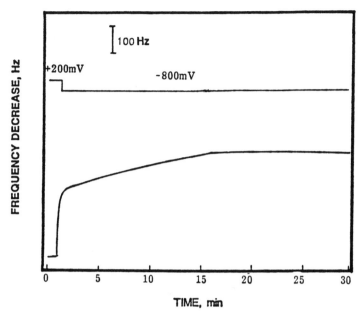

Figure 5a. The frequency change-time transient obtained for chlorpromazine loading into an EQCM, PPy/Melanin film from 0.01 M chlorpromazine solution upon the potential step application from E = 200 mV to E = -800 mV vs. SCE.

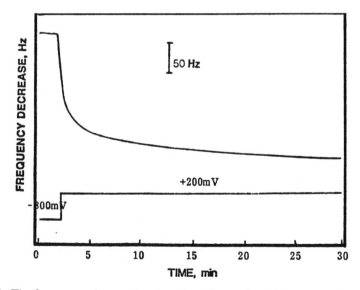

Figure 5b. The frequency change-time transient illustrating Chlorpromazine release from a PPy/Melanin film loaded with chlorpromazine into 0.1 M NaCl solution upon the potential step application from E = -800 mV to E = 200 mV.

The rate of CPZ release strongly depended on the potential at which the release was being carried out. This dependence is illustrated in Figure 6 for step-wise potential changes form E = -800 mV to E = +400 mV. The mass decrease due to the CPZ ejection was much faster at higher anodic potentials, as the slope dm/dt becomes steeper. This is of great importance for controlled drug release, because it allows one to program both the speed and dose of the drug to be released by changing the height and duration of the potential step.

Incorporation of Thioridazine into a PPy/Mel Polymer Composite Film

Thioridazine is a drug from the phenothiazine group which is similar to Chlorpromazine. Thioridazine forms monocations in neutral aqueous solutions, therefore, it would be expected that Thioridazine would be accumulated in the PPy/Mel matrix upon reduction. However, the size of Thioridazine is larger than Chlorpromazine, due to its longer side chain. It was difficult to predict whether this increase in size could cause a significant hinderance to the incorporation of Thioridazine into the PPy/Melanin matrix. However, since the side chain can rotate freely, the increase in length may not be that important if only geometric considerations are taken into account. On the other hand, the change in physiological properties and

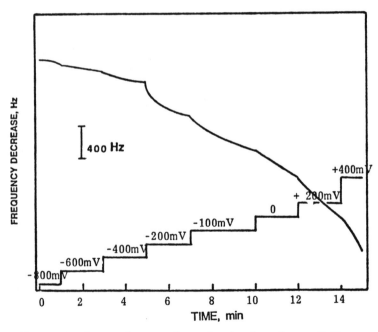

Figure 6. Frequency change-time transients showing the release of chlorpromazine into 0.1 M NaCl solution from a composite PPy/Melanin film loaded with CPZ, upon many subsequent potential steps changing from E = -800 mV to +400 mV vs. SCE.

Thioridazine HCl

most probably in the strength of interactions of the drug molecule with melanin suggests that this relatively small structural difference between the Chlorpromazine and the Thioridazine may not be neglected. In addition to differences in the side chain of these two molecules, there are differences in the substituent groups in the 2-position. The Thioridazine has a $-SCH_3$ substituent in the 2-position, which is responsible for the increase of "antiaromatic character" of the molecule. This group lowers the stability of the radicals and activates the sulfur in the thiazine system in the 2-position altering the electrophilic reaction. In order to evaluate the effect of the side chain on the phenothiazine drug ingress into the PPy/Mel matrix, a series of experiments was performed using a potential step method.

Presented in Figure 7a is the frequency change - time transient obtained for an EQCM PPy/Melanin film in 0.01 M Thioridazine solution for a potential step from E = +200 mV to E = -600 mV vs. SCE. A sharp frequency decrease corresponding to the mass increase was observed. This indicates that the presence of longer side chain does not affect the ability of the phenothiazine drug to penetrate into the PPy/Melanin film upon reduction. Subsequent release of the Thioridazine from the PPy/Mel film was also studied. The frequency change - time transient, shown in Figure 7b, illustrates the dynamics of the Thioridazine release which is forced by a potential step into the PPy oxidation region, from E = -600 mV to E = +200 mV. The initial slope of the transient curve, after the potential step, is somewhat lower than that recorded during the Thioridazine uptake. However, the rate of the Thioridazine egress is sufficiently high to guarantee full reversibility of the accumulation/release process in the time scale of a few minutes. The time scale will, of course, depend on the film thickness.

The effect of the potential switching on the frequency change (mass change) in 0.01 M Thioridazine solution is presented in Figure 8. The potential was repetitively changed from E = +200 mV to E = -600 mV vs. SCE and back to E = +200 mV with a pulse duration of t = 2 minutes. The frequency decrease (mass increase) corresponding to the incorporation of Thioridazine into polymer matrix was observed at E = -600 mV. The frequency increase (mass decrease) corresponding to the Thioridazine release from the polymer matrix was observed, when potential is stepped back to E = +200 mV.

Dynamics of the Incorporation and Release of Triflupromazine

The specific physiological and neuroleptic properties of the drug Triflupromazine from the phenothiazine group are due to the presence of fluorine substitutes in the

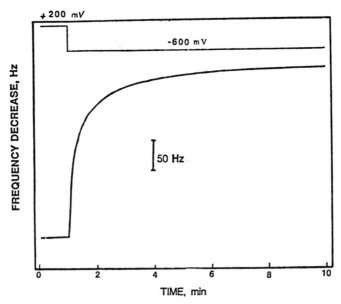

Figure 7a. Frequency change-time transient obtained from Thioridazine loading into a EQCM PPy/Melanin film from 0.01 M Thioridazine solution upon the potential step application from E = + 200 mV to E = -600 mV vs. SCE.

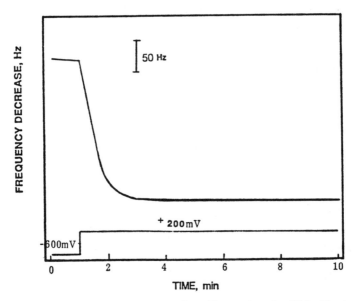

Figure 7b. Frequency change-time transient illustrating the Thioridazine release from a PPy/Melanin/Thioridazine film in 0.01 M NaCl solution upon the potential step application from E = -600 mV to E = +200 mV vs. SCE.

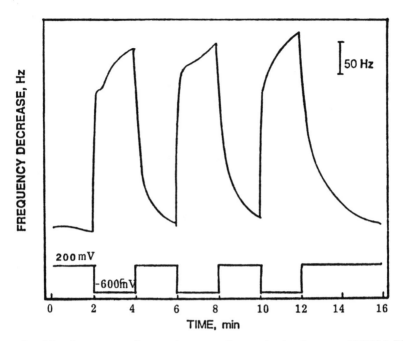

Figure 8. The frequency change-time transients obtained at an EQCM PPy/ Melanin modified gold electrode in 0.01 M Thioridazine solution for potential steps every 2 minutes from E = 200 mV to E = -600 mV vs. SCE.

phenothiazine ring system. Electron-attracting groups, such as the CF_3^- substituent on the Triflupromazine, render the radical formation more difficult and are responsible for the differences in the delocalization of the unpaired electron in the tricyclic system. This structural alteration results in considerable changes in the electron charge distribution within the ring system.

Triflupromazine HCl

To determine whether this charge redistribution can influence the ability of the phenothiazine analog to penetrate into the PPy/Mel polymer matrix, we performed cyclic potential step experiments using the same PPy/Mel composite polymers modified gold EQCM electrode as in previous experiments with other phenothiazine drugs. The results obtained indicate that there is no significant influence by the fluorine atoms on the uptake and release of the drug. The dependence of the amount of the Triflupromazine released from the PPy/Mel film loaded with the drug, during step-wise oxidation of PPy, was similar to other phenothiazines. The temporal evolution of the quartz crystal oscillation frequency during such potential steps from E = -800 mV to E = +200 mV is illustrated in Figure 9. At more anodic potentials, the rate of the Triflupromazine release increased similarly to the other phenotihazines.

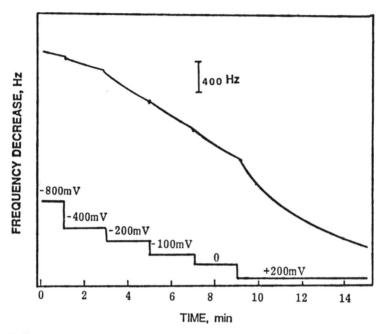

Figure 9. Frequency change-time transients showing the release of Triflupromazine into 0.1 M NaCl from a composite PPy/Melanin film loaded with Triflupromazine upon subsequent potential steps from E = -800 mV to E = +200 mV vs. SCE.

Dynamics of Ingress and Egress of Imipramine with PPy/Melanin Film

Imipramine is a tricyclic antidepressant. The structure of Imipramine resembles that of phenothiazine drugs; the major difference being in the central ring, which is a nonaromatic 7 member ring able to fold and thus adjust to requirements of limited space in a polymer matrix. Therefore, we expected that Imipramine ingress to the PPy/Mel composite films will not be hindered and that this drug could be studied as a potential candidate for our cationic drug release system. The experiments were similar to those reported for the phenothiazine drugs.

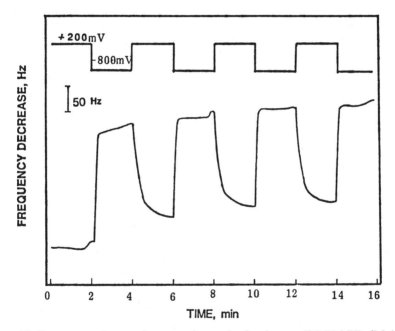

Imipramine (Tofranil)

The reversibility of the ingress and egress processes of Imipramine to and from the PPy/Mel matrix was also investigated using cyclic potential step method. The frequency change - time transients presented in Figure 10 were obtained for a square wave potential program with a potential step between E = +200 mV and E = -800 mV. At the potential E = +200 mV, the PPy undergoes oxidation and the positive charges of polarons and bipolarons force the Imipramine cations out of the film, and this causes the oscillation frequency of the piezoelectrode to increase (the electrode mass decreases). When the potential is switched to E = -800 mV, the PPy backbone is being reduced and the excess negative charge of the melanin drives the Imipramine cations back to the film causing the oscillation

Figure 10. Frequency change-time transients obtained at an EQCM PPy/Melanin modified gold electrode for potential steps every 2 minutes from E = +200 mV to E = -800 mV in 0.01 M Imipramine solution.

frequency to decrease (the electrode mass increases). The rate of the Imipramine release is slightly lower than the rate of the uptake. At the 2-minute potential pulse duration, a slow steady increase in the electrode mass is observed during the first 5 potential cycles, and after that the mass relaxation curves bounce between two well defined limits corresponding to the electrode state at the two potentials.

When a faster switching potential was applied to the PPy/Mel piezoelectrode in Imipramine solution, a triangular frequency vs. time response was obtained, as shown in Figure 11 for the pulse width of 6 seconds. Clearly, neither the ingress nor the egress can be completed during this short pulse duration. Under these conditions, a complex concentration profile of the Imipramine counterions in the polymer film developed, and after a few potential cycles the oscillator frequency (and mass) relaxation became almost steady. A small background noise indicated that the system has a tendency to increase or decrease the average total concentration of the drug in the polymer film.

Conclusions

A new type of composite polypyrrole film has been synthesized. The incorporation of bulky molecules like melanin during the polymerization process of pyrrole

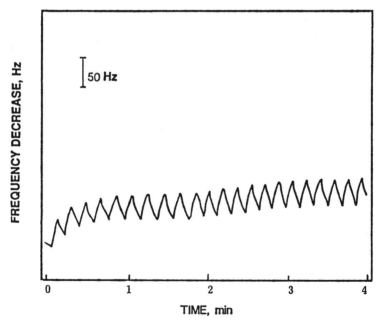

Figure 11. Frequency change-time transients obtained at an EQCM PPy/Melanin modified gold electrode for potential steps every 2 seconds from E = +200 mV to E = -800 mV in a 0.01 M Imipramine solution.

resulted in films with cation-exchange properties. These dopants were permanently incorporated into polypyrrole and were not released. However, the release of cationic drugs from this film was affected by electrochemical stimulation. These films have the ability to incorporate cationic neuroleptic and tricyclic antidepressant drugs at negative potentials, *e.g.*, -800 mV vs. SCE and release them by application of potential step to positive potentials, *e.g.*, + 200 mV vs. SCE.

The electrochemical quartz crystal microbalance (EQCM) was used as a sensitive piezoelectric sensor for the controlled drug release studies. Additional evidence for the release of drugs from polymer film was obtained from UV-Visible spectroscopy data. The EQCM was found to be a convenient technique in studies of controlled drug release from the polymeric matrix.

In conventional therapy, the entire body is subjected to a drug when only a small local concentration is required. These new composite polymers could be applied to controlled release of drugs to the CNS. A controlled release dosage of neuroleptic drugs at a specific site could provide the desirable level of drug and at the same time could provide more consistent protection without toxic effect. An application of these polymers for transdermal delivery of these drugs has also been considered.

Acknowledgements. This work was partially supported by the Bristol-Myers Squibb Company Award of Research Corporation and FDP Grant, Potsdam College, SUNY. Help from undergraduate students F. Mahdavi, L. Dentrone, and E. Seymour is acknowledged.

Literature Cited

1. Langer, R. *Pharmac. Ther.* **1983**, 21, 35-51.
2. *Polymeric Drugs and Drug Delivery Systems.* Dunn, R.L., and Ottenbrite, R.M., Eds., ACS Symposium Series No. 469, 1991.
3. Kydonieus, A. *Treatise on Controlled Drug Delivery.* Marcel Dekker, Inc., New York, 1991.
4. Denizli, A., Kiremitci, M., and Piskin, E. *Biomaterials* **1988**, 9, 257-262.
5. Ottenbrite, R.M. *Encyclopedia of Polymer Science and Engineering,* Suppl. Volume, John Wiley & Sons Inc., New York, 1989.
6. *Polymers for Controlled Drug Delivery,* Tarcha, P.J., Ed., CRC Press, Inc., Boca Raton, Florida, 1990.
7. Graham, N.B., and Wood, D.A. Chapter 8 in *Macromolecular Biomaterials,* Hastings, G.W., and Ducheyne, P., Eds. CRC Press, Inc., Boca Raton, Florida, 1984, p. 181-214.
8. Urquhart, J., Fara,J.W., and Willis, L.W. *Ann. Rev. Pharmacol. Toxicol.* **1984**, 24, 199-263.
9. Lee, P.I., and Good, W. R. Chapter 1 in *Controlled-Release Technology, Pharmaceutical Applications,* Lee, P.L., and Good, W.R. Eds. ACS Symposium Series 348, American Chemical Society, Washington, D.C., 1987.

10. *Anionic Polymeric Drugs*, Donaruma, L. G., Ottenbrite, R. M., and Vogl, O., Eds., John Wiley & Sons, New York, 1980.

11. Pharriss, B. B. *J. Reprod. Med.* **1978**, 20, 155-65.

12. Goldman, P. *Arch. Ophthalmol.* **1982**, 93, 74-86.

13. Spangler, D.L., Kalof, D.D., Bloom, Fl.L., Wittig, H.J. *Ann. Allergy* **1978**, 40, 6-11.

14. Theeuwes, F. *Curr. Med. Res. Opin.* **1983**, 8 (Suppl. 2), 20-27.

15. Mroczek, W. J., Ulrych, M., Yoder, S. *Clin. Pharmacol. Ther.* **1982**, 31, 252 (Abstr.)

16. Muller, P., Imhof, P.R., Burkart, F., Chu, L.-C., Gerardin, A. *Eur. J. Clin. Pharmacol.* **1982**, 22, 473-80.

17. Shaw, J., Urquhart, J. *Trends Pharmacol. Sci.* **1980**, 1, 208-11.

18. Kaplan, A.M., Kuss, K., and Ottenbrite, R. M. *Ann, NY Acad. Sci.* **1985**, 446, 169.

20. Graham, N.B. *British Polymer Journal* **1978**, 10, 260-266.

21. Ormrod, D.J., Cawley, S., and Miller, T.E. *Int. J. Immunopharm.* **1985**, 7, 443-448.

22. Yamada, A., Sakurai, Y., Nakamura, K., Hanyu, F., Yoshida, M., and Kaetsu, I. *Trans. Am. Soc. Artif. Intern. Organs* **1980**, 26, 514-517.

23. Dunkan, R., Seymour, L. W., Ulbrich, K., and Kopecek, J. *J. Bioactive and Compatible Polymers* **1988**, 3, 4-15.

24. Rosen, H.B., Chang, J., Wnek, G.E., Linhardt, R.J., and Langer, R. *Biomaterials* **1983**, 4, 131-133.

25. Heller, J., and Trescony, P.V. *J. Pharm. Sci.* **1979**, 68(7), 918-921.

26. Fischel-Ghodsian, F., Brown, L., Mathiowitz, E., Brandenburg, D., and Langer, R. *Proc. Natl. Acad. Sci. USA* **1988**, 85, 2403-2406.

27. Hsieh, D.S.T., Langer, R., and Folkman, J. *Proc. Natl. Acad. Sci. USA* **1981**, 78(3), 1863-1867.

28. Maeda, H., Oda, T., Matsumura, Y., and Kimura, M. *J. Bioact. Compatible Polymers* **1988**, 3, 27-43.

29. Kopecek, J. *J. Bioact. Compatible Polymers* **1988**, 3, 4.

30. Duncan, R., Seymour, L.W., Ulbrich, K., and Kopecek, J. *J. Bioact. Compatible Polymers* **1988**, 3, 4.

31. *Bioreversible Carriers in Drug Design Theory and Application*, Roche, E. B., Ed., Pergamon Press, Elmsford, NY, 1987.

32. Costall, B., Domeney, A.M., Naylor, R.J. *Br. J. Pharmacol.* **1981**, 74, 899P-900P (Abstr.)

33. Bodor, N., Brewster, M.E. *Pharmacol. Ther.* **1983**, 19, 337-386.

34. Siew, C., Goldstein, D.B. *J. Pharmacol. Exp. Ther.* **1978**, 204, 541-46.

35. Kroin, Y.S., Penn, R.D. *Neurosurgery* **1982**, 10, 349-354.

36. de la Torre, J. C., Gonzalez-Carvajal, M. *Lab. Anion. Sci.* **1981**, 31, 701-703.

37. Hepel, M. *Proceedings of the Third International Meeting on Chemical Sensors*, Cleveland, Ohio, September 24-26, 1990, pp. 35-37.

38. Hepel, M. in: *Proceedings of the International Symposium on Composites, Am. Cer. Society,* Orlando, Florida. Ceramic Transactions "Advanced Composite Materials" (Ed.) Sacks, M. Vol. 19, 1991, pp. 389-396.

39. Zhou, Q.X., Miller, L.L., and Valentine, J.R. *J. Electroanal. Chem.* **1989**, 261, 147-164.
40. Miller, L.L., and Zhou, Z. X. *Macromolecules* **1987**, 20, 1594-1597.
41. Hepel, M., Kanige, K., and Bruckenstein, K. *Langmuir* **1990**, 6, 1063-1067.
42. Hepel, M., Seymour, E., Yoglev, D., and Fendler, J. *Chem. Materials* **1992**, 4, 209-216.
43. Naoi, K., Lien, M., and Smyrl, W.H. *J. Electrochem. Soc.* **1991**, 138, 440.
44. Borjas, R., and Buttry, D.A. *Chem. Mater.* **1991**, 3, 872.
45. Hillman, A.R., Swann, M. J., and Bruckenstein, S. *J. Phys. Chem.* **1991**, 95, 3271.
46. Ward, M. D. *J. Phys. Chem.* **1988**, 92, 2049.
47. Qiu, Y. J., Reynolds, J. R. *Polym. Eng. and Sci.* **1991**, 31, 417.

RECEIVED May 6, 1993

Chapter 8

Pulsatile Drug Release by Electric Stimulus

You Han Bae, Ick Chan Kwon, and Sung Wan Kim

Department of Pharmaceutics and Pharmaceutical Chemistry, Center
for Controlled Chemical Delivery, University of Utah, Salt Lake
City, UT 84112

Electric currents have been applied to polymeric monolithic devices
to produce pulsatile drug release. The polymeric matrices used were
either electrically charged networks, or polymer-polymer complexes
based on hydrogen bonding or electrostatic interactions. Positively
charged drug was released in an on-off manner from a negatively
charged network. The ionically bound drug was freed from the
polymer chains by ion-exchange with H^+ ions. The polymer-polymer
interactions in the formed complexes were perturbed by ionization or
deionization of one part of the polymer pairs by either increasing or
decreasing pH around electrodes. This resulted in polymer surface
erosion and pulsatile release of the drugs entrapped in the matrices.

There are many clinical situations where pulsatile drug administration, drug release
responding to pathological conditions, or drug release matching circadian rhythms
are more effective than a constant drug release (1,2). Such situations include
cancer chemotherapy, insulin treatment of diabetes, general hormone replacement,
birth control, delivery of gastric acid inhibitor for ulcer treatment, antiarrhythmic
delivery for heart rhythm disorder, and immunization (3).

To produce such drug release patterns, an externally modulated (open-loop
system) or self-regulated drug delivery (closed-loop system) using bioengineering,
biochemical, biological or polymer-related technologies must be designed.

In this context, stimuli sensitive hydrogels are applicable for modulated drug
delivery because these polymers respond to external stimuli to control the release
rate of drugs (3-5). External signals used in a pulsatile drug release system are
ultrasound (6,7), temperature (8,9), magnetic field (10), photo irradiation (11) and
electric field (12-18). From a practical point of view, one of the most convenient
signals for use in pulsatile drug delivery would be an electric current.

Electrically modulated solute release has been demonstrated by several

0097–6156/94/0545–0098$08.00/0

methods. These include electrically controlled resistance to transport of ions across polypyrrole membranes *(19)*, cathodic current induced cleavage of chemically bound drug via amide linkage *(20)*, and polyelectrolyte gel shrinkage under electric current, resulting in "squeezing out" of imbibed drug solutions *(21)*.

One effect of the electrochemical reactions in aqueous system is the local pH changes around electrodes. Through water electrolysis, hydronium ions (H^+) are generated at the anode, while hydroxyl ions (OH^-) are produced at the cathode. These local pH changes were utilized for controlling 'on-off' solute release via ion exchange (Figure 1) *(22)* or erosion of interpolymer complex gels (Figure 2) *(23,24)*. This paper summarizes these two phenomena.

Experimental Methods

Synthesis

Polyelectrolyte Gel. A crosslinked random copolymer of 2-acrylamido-2-methyl-1-propane sulfonic acid (AMPS)/n-butylmethacrylate (BMA) was synthesized using ethylene glycol dimethacrylate as a crosslinker (0.8 mol%) and N,N'-azobis-isobutyronirile as an initiator (0.1 mol%). The feed composition for AMPS/BMA was 27/73 mole ratio. A mixture of monomer (5.39 g), solvent (DMF; 5 mL), initiator and crosslinker was bubbled with dried nitrogen gas for 20 min to remove oxygen. Polymerization was performed between two polyester plates, separated by a silicone rubber gasket (1 mm diameter), or in a glass test tube (10 mm diameter, 100 mm length) at 60°C for three days. After polymerization, the polymers were removed from the mold or test tube and soaked for 1 week in a water/acetone (50/50 v/v%) solvent mixture to extract unreacted compounds (daily solvent change). After extraction, the polymers were stored in distilled-deionized water (DDW) until use.

Interpolymer Complexes. An insoluble polymer complex was formed between two water soluble polymers, poly(ethyloxazoline) (PEOx) and poly(methacrylic acid) (PMAA). PEOx and PMAA form hydrogen bonded complexes between the carboxylic groups in PMAA and the repeating units of PEOx.

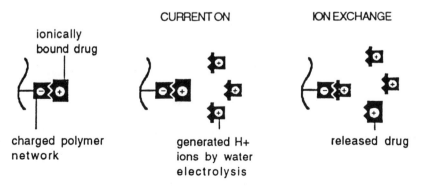

Figure 1. Ion exchange of positively charged drug bound to negatively charged polymer network with hydrogen ion generated by water electrolysis by electric current.

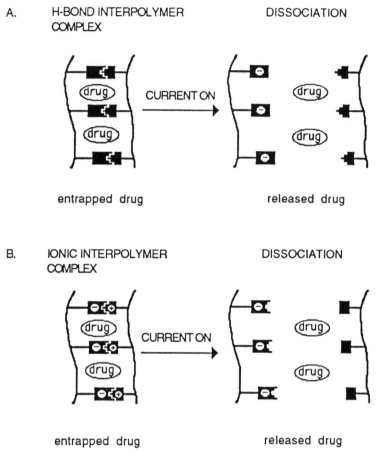

Figure 2. Schematic presentation of erodible polymer matrix by an electric current.

Aqueous solutions of PEOx (0.099 g, average Mw 500,000) and PMAA (0.086 g, average Mw 60,000) were made by dissolving each polymer in 10 mL DDW. The complex was formed by adjusting the pH of the solution to pH 5. The complex was filtered and then swollen for 1 hour in acetone/water (65/35 v/v). The swollen sticky polymer complex was placed between two Teflon blocks to from a disk. The molded, disk shaped matrix was 2 mm thick and 15 mm in diameter and was dried under vacuum for 3 days.

Another interpolymer complex was formed using poly(allyl amine) (PAA) (cationic polymer at neutral pH) and a bioactive polyanion, heparin. PAA (average Mw 60,000) and low molecular weight heparin (average Mw 6,000) (50 mg each) were dissolved separately in 10 mL DDW. The polymer complex immediately formed upon mixing the two polymer solutions. The complex was filtered, washed with DDW, and vacuum dried overnight. The dried complex was ground into a fine powder and pressed into a disk matrix (13 mm diameter) with a press (6 ton/cm^2 pressure).

Swelling Measurement

Polyelectrolyte Gel. The membrane, equilibrated in DDW at room temperature, was cut into 14 mm diameter disks (1.5 mm thickness) and dried under vacuum at room temperature until no detectable weight change was evident. The dried disks were placed in KCl/HCl buffer media (10 different pH values; range 0.68-2.25) *(25)* until equilibrium was reached; generally, within three days. The degree of swelling was calculated by measuring the weight of swollen disks until weight changes were within 1% of the previous measurement. The degree of swelling was expressed as a ratio of absorbed water weight to dried polymer weight (W_w/W_p).

Interpolymer Complexes: Equilibrium swelling of PEOx/PMAA and PAA/heparin disks were measured in the 0.9% saline solution (pH 5.1 $\pm$ 0.15) and isotonic phosphate buffered saline (PBS, pH 7.4) respectively. The equilibrium swelling levels of the disks (0.43 for PEOx/PMAA complex and 0.54 for PAA/heparin complex) were reached within 3 days.

Solute Loading

Polyelectrolyte Gel. For solute release experiments, two model compounds were loaded into crosslinked random copolymers of AMPS/BMA and released in DDW under an applied electric current; hydrocortisone (nonionic solute) and edrophonium chloride (a positively charged solute).

The dried poly(AMPS/BMA) gel (59.3 mg) was equilibrated in 50 mL of a saturated solution of hydrocortisone in ethanol for three days. The drug loaded swollen gel was air dried in a half-covered 100 mL beaker for one week at room temperature and then vacuum dried overnight at 25°C (76.6 mg after drying process).

Edrophonium chloride was loaded into the gel by an ion-exchange method. The dried poly(AMPS/BMA) gel (42.3 mg) was equilibrated for 3 days in 50 mL aqueous solution of 0.5 N NaOH to convert the -SO_3H groups to -SO_3Na and then the -SO_3Na groups were changed to -SO_3Ed (where EdCl is edrophonium chloride) by immersing the gel in a saturated aqueous solution of edrophonium chloride. Physically entrapped solute was washed out completely by soaking in DDW until no further detection of solute in water at 273 nm was observed. Edrophonium chloride loaded gels were vacuum dried at 50°C for one week (52 mg weight after drying process).

Interpolymer Complexes. Insulin was selected to investigate the PEOx/PMAA complex system for modulated solute release. Regular zinc insulin (10 mg) was suspended in 10 mL of the mixed polymer solution (pH 5.5), and the complex was formed by decreasing the pH below pH 5 with 0.1 N HCl solution. The insulin containing polymeric complex was then filtered, dried, and compressed into a disk as previously described.

Release Experiments

Polyelectrolyte Gel. Release experiments for edrophonium chloride were conducted in a 20 mL glass bottle filled with DDW and equipped with a magnetic stirrer. The solute loaded gel disk (14 mm diameter, 1.5 mm thickness) was attached to platinum electrodes (1 cm gap). Various electric fields (3-6 V/cm) were applied in a step function at 30 min intervals. The release medium was replaced every 10 min during the application of electric fields.

For hydrocortisone release, the dried, solute loaded gel was immersed in 100 mL DDW. The released amount of hydrocortisone was measured at 1 hour intervals in 100 mL of water in the absence of electric current and 15 min intervals in 20 mL of water when an electric field (8 V/cm) was applied. The amount released was measured by UV spectroscopy (Perkin-Elmers Lambda 7 UV/VIS Spectrophotometer) at 242 nm for hydrocortisone and 273 nm for edrophonium chloride.

Interpolymer Complexes. To investigate solute release from interpolymer complexes with an electric current, the swollen matrix was attached to a woven platinum cathode in a continuously stirred release media (0.9% saline solution for PEOx/PMAA complex and isotonic PBS (pH 7.4) for PAA/heparin complex) and an electric current was applied (see Figure 3). The amount of insulin released was determined by radioimmuno-assay (RIA) method. For heparin assay, the Azure II method was used at pH 11, to prevent complexation of the two dissolved polymers.

Results and Discussion

Polyelectrolyte Gel (Ion Exchange). The loading content of ionically bound

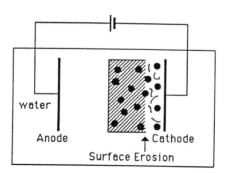

Figure 3. Schematic illustration of solute release experiment from electro-erodible polymers.

edrophonium chloride in poly(AMPS/BMA) matrix was 18 wt% (97% of -SO$_3$H was exchanged to -SO$_3$Ed, theoretically).

The release of edrophonium chloride from this matrix with varying intensity of electric stimulation in DDW shows a complete "on-off" release profile, as shown in Figure 4. The overall release rate decreased with time and this decrease was a function of an increased electric field. This may have been caused by the depletion of the loaded solute during the application of electric field.

To deduce the relationship between applied electric current and solute release rate, the electric current during the application of a constant electric potential was monitored. Since the current increased during the application of a constant electric potential, the average current value during the first cycle was obtained and plotted versus the average release rate of the first cycle, which is less affected by the amount of loaded solute (Figure 5). The release rate increased with an increase of electric current and the magnitude of the release rate was regulated by the electric current.

This release pattern can be explained by following schemes. In the first step, ion exchange between a positively charged solute and hydrogen ions produced at the anode by water electrolysis takes place at the site of -SO$_3$Ed groups along the polymer. In second step, the released, positively charged solute migrates to the

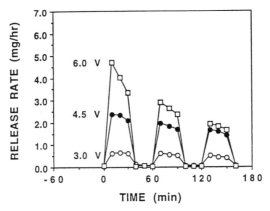

Figure 4. Edrophonium chloride release from poly(AMPS/BMA) gel in distilled-deionized water with various pulsatile electric current.

cathode and diffuses out of the membrane. The release of solute is enhanced by the squeezing effect at the anode side and by electro-osmosis. Since all physically entrapped solute was extracted by washing with DDW, only ionically bound solute exists inside the gel. Thus, it is assumed that the exchange rate is controlled by the electric current. Therefore, the rate determining step for positively charge solute release is ion exchange, followed by rapid release of solute from the device.

Changes in pH under an applied electric current will affect swelling of poly(AMPS/BMA) gel. The swelling properties of crosslinked poly(AMPS/BMA)

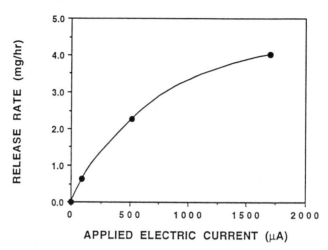

Figure 5. Effect of electric current on edrophonium chloride release from poly(AMPS/BMA) gel in distilled-deionized water. (Reprinted with permission from ref. 22. Copyright 1991 Elsevior Science Publishers BV.)

as a function of pH are shown in Figure 6. The maximum swelling change of this gel occurred at pH 1-1.5 which may reflect the ionization of the AMPS sulfonic acid groups. The degree of poly(AMPS/BMA) swelling was controlled by changes in the pH of the medium in the range of pH 1-2.5. Above pH 2.5, gel swelling increased slightly, indicating that most acidic groups were ionized (90% theoretical ionization at pH 2.5, if the pK_a of polymer gel is 1.5 from the inflection point in Figure 6).

However, by applying an electric current to the gel immersed in DDW by electrodes in contact with the gel, deswelling occurred at the anode side while no dimensional changes were visually observed at the cathode side. This anisotropic deswelling may have been caused by an electrochemical process, although, in other studies *(16,26,27)*, explanations for the mechanism of swelling changes of polyelectrolyte gel under an electric field vary from case to case. Hydrogen ions generated by water electrolysis at the anode reduce the pH near the electrode, resulting in local deswelling. The hydroxyl ions produced by the cathode should increase the local pH, but increased pH did not affect the swelling behavior of poly(AMPS/BMA) significantly. The pH inside the gel, equilibrated with DDW before the application of electric current, was 2.4, measured by inserting pH microelectrode (Model MI 415, Microelectrodes Inc., Londonderry, NH). The change of swelling properties of this gel was not significant at a pH above 2.5 since most acidic groups were already ionized. Therefore, it is obvious that the local pH changes near the electrodes provide a reasonable explanation for anisotropic deswelling behavior under an electric current.

To examine this deswelling effect on solute release, a neutral solute, hydrocortisone, was released under an applied electric field. Since this solute is

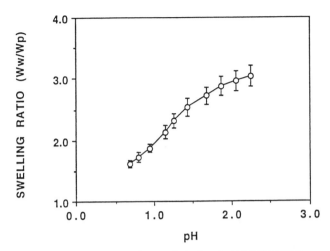

Figure 6. Swelling ratio (W_w/W_p) of crosslinked poly(AMPS/BMA) as a function of pH. (Reprinted with permission from ref. 22. Copyright 1991 Elsevior Science Publishers BV.)

physically entrapped inside the gel (23 wt% loading), the change of release pattern could only be affected by swelling changes in the gel.

Hydrocortisone release in DDW water from the initially dried gel with stimulation of electric current is shown in Figure 7. The overall release rate decreased with time and showed pulsatile release when electric current was applied. These pulses may be attributed to the bulk squeezing caused by anisotropic deswelling under the electric field. However, the amount of released hydrocortisone was one order of magnitude less than that of positively charged edrophonium chloride. Therefore, the effect of squeezing on the release of edrophonium chloride is a minor factor, but squeezing can enhance the release of edrophonium chloride exchanged with the hydrogen ions.

In the deswelling process in air, water dripping during the application of electric current was found only at the cathode side of the gel *(22)*. A similar experiment was done by Osada *et al. (19)* and they observed gel movement towards the anode and water dripping near the cathode. In their experiment, poly(AMPS) gel was located between fixed electrodes. They suggested electrophoretic movement of the gel to explain the observed result. However, our results indicate that deswelling in air under an electric field may not be caused by an electrophoretic phenomenon of the negatively charged network, since no considerable deswelling or gel movement was observed when the experiment was conducted in water as described before.

One possible explanation for water dripping at the cathode side in air could be electro-osmosis resulting in water movement inside the gel. This explanation is supported by the experiment with the two chamber diffusion cell which showed a

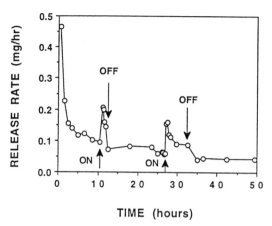

TIME (hours)

Figure 7. Hydrocortisone release from poly(AMPS/BMA) gel in distilled-deionized water with a pulsatile electric field (8 V/cm).

pumping of water from the anode to the cathode side through the gel membrane under applied electric currents *(22)*.

Therefore, the deswelling at the anode side in the absence of water medium was affected not only by pH changes, but also by electro-osmosis resulting in water dripping at the cathode side. On the other hand, the deswelling behavior of the gel with applied electric current in water is different from that in air because of the continuous water supply from the environment.

In summary, the "on-off" release pattern for edrophonium chloride was attributed to ion exchange of the positively charged solute with hydrogen ions concentrated at the anode side. Three phenomena were attributed to the rapid "on-off" response under an electric current application: ion exchange, deswelling at the anode, and electro-osmosis from the anode to the cathode. These phenomena were enhanced by increasing electric field strength. The mechanism for the controlled release of physically entrapped neutral solute (hydrocortisone) inside the gel was based on passive diffusion plus gel squeezing caused by deswelling at the anode side.

Interpolymer Complex (Erosion). PEOx and either PMAA or poly(acrylic acid) form complexes via intermolecular hydrogen bonding. There are two possible sites in the repeating unit of PEOx for hydrogen bonding with carboxylic groups: the carbonyl oxygen and the nitrogen. The carbonyl oxygens are probably the dominant interactive group due to a 1:1 ratio of repeat units (28).

The complex was formed below pH 5 and dissociated above pH 5.4. This was attributed to the deionization and ionization of the carboxylic groups of PMAA with pH changes. The discrepancy between the precipitation and dissolution pH values may result from a change in pKa for the PMAA carboxylic groups before and after complex formation or cooperative interaction in complex formation.

During the application of an electric current, the solid matrix surface facing the cathode began to dissolve. Since the cathode produces hydroxyl ions (electrolysis), the local pH near the cathode increased and hydrogen bonding between the two polymers was disrupted, resulting in disintegration of the polymer complex into two water soluble polymers. The rate of polymer weight loss was constant until 80% of the initial disk weight was lost. The linearity of weight loss vs. time implies that this process occurs through erosion of the cathode facing surface. A stepwise weight loss was also observed by applying an electric current (10 mA) in a step function until 80 % total weight loss was reached as shown in Figure 8.

The loading content of insulin in the PEOx/PMAA complex was 0.5 ± 0.2 wt% (n = 3). The disks were soaked in 0.9 % saline solution for 3 days before applying the electric current. The amount of insulin released during the 3 day equilibrium period was less than 4% of initial insulin load.

By applying a step function of electric current to the insulin loaded matrix, insulin was released in an "on-off" manner until 70 % of the loaded insulin was released, as demonstrated in Figure 9. The large deviation of the release rate was from irregular erosion of the device, attributed to defects in the insulin loaded device. The release of dispersed insulin in the matrix was hardly detectable in the absence of electric current and the insulin release was a function of the surface erosion rate of polymer.

Although the system based on hydrogen bonding complex demonstrated a well defined insulin release in an "on-off" manner by electric current, the working condition was limited to acidic pH due to the nature of carboxylic group. For a system which can work at isotonic neutral pH, a polyelectrolyte complex was tested.

Heparin, a bioactive polyanion, was complexed with PAA. At neutral pH,

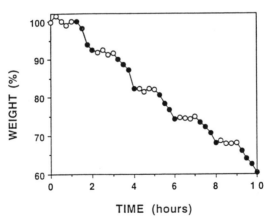

Figure 8. Weight loss of poly(ethyloxazoline)/poly(methacrylic acid) complex matrix in 0.9 % saline solution when 10 mA step function electric current was applied: ● - current on; ○ - current off. (Reprinted with permission from ref. 23. Copyright 1991 Macmillan.)

POLYMERIC DRUGS AND DRUG ADMINISTRATION

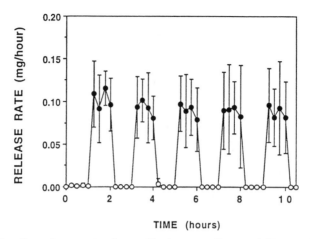

TIME (hours)

Figure 9. Insulin release rate (normalized to mg/h per 160 mg device) from insulin-loaded matrix of poly(2-ethyl-2-oxazoline)/poly(methacrylic acid) complex with the application of step-function electric current in 0.9 % saline solution (mean ± S.D., n = 3): ● - current on (5 mA), ○ - current off. (Reprinted with permission from ref. 23. Copyright 1991 Macmillan.)

the positively charged amine groups in PAA complexed with $-COO^-$ (from iduronic acid and glucuronic acid units) or $-SO_3^-$ (from sulfoiduronic acid and 2,6-disulfoglucosamine units) in heparin. This complex was formed in a wide range of pH from 3 to pH 10, and the resulting complex dissociated below pH 2 (deionization of carboxylic and sulfate groups) and above pH 11 (deionization of amine groups).

During the application of electric current, the solid matrix dissolved from the cathode facing surface due to an increase of pH. This phenomenon was similar to the previous hydrogen bonding complex system, except for a slower dissociation at a given electric current than that of the hydrogen bonding system. The release pattern of heparin showed a complete "on-off" profile in response to the applied electric field, as shown in Figure 10. The release rate was dependent on the intensity of the applied electric current, but, was not linearly proportional to the applied electric current.

Conclusion

An "on-off" release of a positively charged solute from negatively charged hydrogels was demonstrated by applying an electric current. The ion exchange mechanism was facilitated through hydronium ions generated at the anode under an electric current.

Another "on-off" release method was investigated through surface erosion of a polymer complex stimulated by an electrical current. The erosion was either

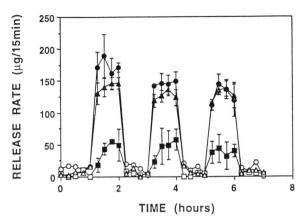

Figure 10. Heparin release rate (normalized to μg/15 min/50 mg device) from heparin-poly(allylamine) complex matrix with the application of step function electric current in PBS 7.4 solution (mean ± S.D., n = 3): ○,□, and △ off state; ● 20 mA, ■15 mA, ▲10 mA.

caused by the disruption of the hydrogen bonded complex by ionization of proton donor or deionization of one of a polymer complex pair via an electrochemically induced local pH change.   In these systems, the drugs can be charged polymeric solutes which are directly involved in polymer complexation or neutral solutes.

Literature Cited

1. Mazer, N. A. *J. Control. Rel.* **1990**, 11, 343-356.
2. Lemmer, B. *J. Control. Rel.* **1991**, 16, 63-74.
3. Kost, J. *Pulsed and Self-Regulated Drug Delivery*, CRC Press, Boca Raton, FL, 1990
4. Kwon, I.C., Bae, Y.H., Okano, T., Berner, B., and Kim, S.W. *Makromol. Chem., Macromol. Symp* **1990**, 33, 265-277.
5. Sawahata, K., Hara, M., Yasunaga, H., and Osada, Y. *J. Control. Rel.* **1990**, 14, 253-262.
6. Okahata, Y., and Noguchi, H. *Chem. Lett.* **1983**, 1517-1520.
7. Kost, J., Leong, K.W., and Langer, R. *Proceed. Intern. Sym. Control. Rel. Bioact. Mater.* **1983**, 10, 84-85.
8. Bae, Y.H.. Okano, T., Hsu, R., and Kim, S.W. *Makromol. Chem., Rapid. Commun.* **1987**, 8, 481-485.
9. Hoffman, A.S., Afrassiabi, A., and Dong, A.C. *J. Control. Rel.* **1986**, 4, 213-222.
10. Hsieh, D.S.T., Langer, R., and Folkman, J. *Proc. Natl. Acad. Sci. USA* **1981**, 78, 1863-1867.
11. Ishihara, K., Hamada, N., Kato, S., Shinohara, I. *J. Polym. Sci.: Polym. Chem. Ed.* **1984**, 22, 881-884.
12. Burgmayer, P., and Murray, R.W. *J. Am. Chem. Soc.* **1978**, 104, 6139-6140.

13. Osada, Y., and Takeuchi, Y. *J. Polym. Sci.: Polym. Lett. Ed.* **1981**, 19, 303-308.
14. Osada, Y., and Takeuchi, Y. *Polym. J.* **1983**, 15, 279-284.
15. Eisenberg, S.R., and Grodzinsky, A.J. *J. Membr. Sci.* **1984**, 19, 173-194.
16. Grodzinsky, A.J., and Weiss, A.M. *Separation and Purification Methods* **1985**, 14, 1-40.
17. Weiss, A.M., Grodzinsky, A.J., and Yarmush, M.L. *AIChE Symposium Series* **1986**, 82, 85-98.
18. Umezawa, K., and Osada, K. *Chem. Lett.* **1987**, 1795-1798.
19. Burgmayer, P., and Murry, R. W. *J. Am. Chem. Soc.* **1982**, 104, 6139-6140.
20. Lau, A.N.K., and Miller, L.L. *J. Am. Chem. Soc.* **1983**, 105, 5217-5277.
21. Sawahata, K., Hara, M., Yasunaga, H., and Osada, Y. *J. Contol. Rel.* **1990**, 14, 253-262.
22. Kwon, I.C., Bae, Y.H., Okano, T., and Kim, S.W. *J. Control. Rel.* **1991**, 17, 149-156.
23. Kwon, I.C., Bae, Y.H., and Kim, S.W. *Nature* **1991**, 354, 291-293.
24. Bae, Y.H., Kwon, I.C., Pai, C.M., and Kim, S.W. *Makromol. Chem., Macromol. Symp.* **1993**, 70/71, 173—181.
25. Dawson, R.M.C., Elliott, D.C., Elliott, W.H., and Jones, K.M. *Data for Biochemical Research*, 3 rd Ed., Oxford University Press, N.Y., 1084, p. 426.
26. De Rossi, D., Chiarelli, P., Buzzigoli, G., Domenici, C., and Lazzeri, L. *Trans. Am. Soc. Artif. Intern. Organs.* **1986**, 32, 157-162.
27. Tanaka, T., Nishio, I., Sun, S., and Ueno-Nisho, S. *Science* **1982**, 218, 467-469.
28. Lichkus, A.M., Painter, P.C., and Coleman, M. M. *Macromolecules* **1988**, 21, 2636-2641.

RECEIVED August 11, 1993

Chapter 9

Poly(ethylene oxide)-Based Delivery Systems

Influence of Polymer Molecular Weight and Gel Viscoelastic Behavior on Drug Release Mechanism

A. Apicella[1], B. Cappello[2], M. A. Del Nobile[1], M. I. La Rotonda[2], G. Mensitieri[1], L. Nicolais[1], and S. Seccia[2]

[1]Department of Materials and Production Engineering, University of Naples Federico II, 80125 Naples, Italy
[2]Department of Pharmaceutical and Toxicological Chemistry, University of Naples Federico II, 80131 Naples, Italy

The release kinetics of monolithic drug delivery devices based on poly(ethylene oxide) (PEO) was investigated. Water soluble PEO's with molecular weights of 600,000 and 4,000,000 and two blends were studied. The water-swelling and dissolution characteristics of the two polymers and two blends, along with the diffusivity of the drug etofylline in the water-penetrated polymer gels were analyzed. The higher molecular weight PEO and the two blends underwent swelling rather than polymer dissolution. As a result, a non-constant release induced by the prevailing diffusive control was observed. Conversely, the drug release from the lower molecular weight PEO was a result of equal rates of swelling and dissolution, which gave a constant release rate.

The diffusion controlled release technology based on polymer barrier characteristics is a good alternative to conventional delivery systems. Complex reservoir systems and monolithic matrix systems are two important applications. For the former, a zero order kinetics release may be maintained until the drug activity in the reservoir is kept constant. On the other hand, environmentally passive monolithic matrix systems, which contain dispersed or dissolved active ingredients, are not able to produce a constant delivery rate, at least for devices with simple geometries. In such cases, the diffusion control generally leads to a square root time dependency for the drug release.

The use of environmentally interactive monolithic devices made of hydrophilic polymers were first proposed by Hopfenberg (1,2), and then extensively investigated by several researchers (3-8). As the penetrant enters the drug entrapping matrix, the polymer swells and the active ingredient diffuses out from the swollen part. This relaxation controlled sorption is governed by the solvent concentration at the interface separating the swollen and the unpenetrated

0097–6156/94/0545–0111$08.00/0

polymer. The polymer at the interface relaxes and swells at a constant rate as long as the penetrant concentration at the moving boundary remains constant. Zero order release from this type of device requires a constant surface area and a constant swelling rate of the polymer matrix (limiting Case II sorption), as well as a high diffusivity of the entrapped species. These conditions, generally, can not be maintained for long release times (2), since they strongly depend on the time evolution of the interactions involving polymer, penetrant and solute.

At the early stages of polymer swelling, due to the thinness of the swollen layer, the diffusive conductances of both the solvent toward the unpenetrated core and the drug moving outside are much higher than the swelling rate. A constant delivery is brought about by the constant rate of the polymer relaxation. However, as the swollen layer thickness increases, diffusive conductance is attained, and both swelling and release rates are progressively reduced. Both Hopfenberg (2) and Peppas (3) proposed similar dimensionless parameters to identify zero order release in terms of initial solvent penetration rate, swelling thickness, and diffusivity of the active ingredient in the swollen polymer layer. In order to provide zero order release from the swelling systems of increasing thickness, the drug should have a high diffusivity in the swollen polymer. To enhance the diffusivity of a drug in the more porous swollen layers, the addition of water soluble components to the polymer has been suggested (4).

Poly(ethylene oxide) (PEO) based systems have been proposed as drug delivery devices. Graham et al. (9-11) reported the constant release of prostaglandin E2 from the crosslinked crystalline-rubbery hydrogel matrices based on PEO. The matrices undergo solvent sorption and crystallites melt in aqueous environments. In the case of uncrosslinked PEO matrices, the solubility of the polymer can alter the characteristics of the penetrated layer. This can lead to different behaviors in the systems with different dissolution features.

In order to control the release of an active agent, there should be a balance between the diffusion of the active agent and the solubilization of the polymer matrix. The diffusivity of the drug through the matrix, the swelling of the polymer and its solubilization rate can be biased by changing the molecular weight of the polymer, or by blending polymer fractions with different molecular weights.

Mucoadhesion is another important property of a polymer used as a matrix for monolithic drug delivery devices. In the development of oral controlled-release devices, bioadhesive polymers may provide a relatively short-term adhesion between the drug delivery system and the epithelial surface of the gastrointestinal tract. Due to the linear flexible structure of the PEO macromolecules, this polymer shows a particular ability to form entangled physical bonds which interpenetrate deeply into mucous networks. The mucoadhesive properties of PEO strongly depend on the polymer molecular weight and are more pronounced in the case of high molecular weight materials (12-14). For example, 20,000 Mw PEO displays no bioadhesion, while 4,000,000 Mw PEO has very good bioadhesion (14).

A theoretical approach to mucoadhesion, based on the scaling of polymer diffusion and relaxation, has been recently presented (15). The fracture energy per unit area, which is related to the adhesion between the two gels (polymer gel and mucus layer), has been theoretically predicted to scale $N^{3/2}$, where N is the degree of polymerization of the macromolecules. As a consequence, low molecular weight polymer gels are expected to develop maximum adhesive strength in shorter time periods even if the fracture energy is lower. The viscoelastic properties of the mucoadhesive polymer gel must also be taken into account. In fact, a good interpenetration between the adjacent layers of the mucus and the polymer gel is ineffective in holding the mucoadhesive tablet at a specific site if the polymer gel does not have good viscoelastic behavior. To limit the tendency of the swollen outer layer of the tablet to flow, it is preferable that the elastic characteristics prevail over the viscous properties. As a consequence, the polymer gel is able to stay in place for a longer time and attain the maximum level of mucoadhesion. To avoid the displacements of the mucoadhesive device due to tissue movements, the polymer gel should also meet shear module requirements.

The water-swelling, mucoadhesive, and viscoelastic properties of a gel strongly depend on the molecular weight of the polymer matrix. Therefore, the release characteristics related to the water swelling behavior, as well as the potential mucoadhesive properties of PEO systems can be finely tuned by blending different molecular weights fractions.

Based on this information, we are interested in using PEO for mucoadhesive control release devices. In the present study, the diffusivity of the active ingredient etofylline in a swollen polymer and the subsequent release behavior were considered in relation to the sorption, melting temperature and dissolution characteristics of two PEO's of different molecular weights.

Release Mechanisms from Monolithic Devices

The release rate of a dissolved or dispersed drug from a polymeric film or tablet in a specific environment depends on the nature of the diffusion and sorption processes which involve the polymer/environment system, as well as the polymer/drug system.

Diffusion Controlled Devices. A dissolved species will diffuse out from a matrix, according to an ordinary diffusion law, if there is no active interaction with the external environment (Figure 1, left). In this case, the concentration profile in the slab decreases with time which leads to a progressive reduction of the release rate (i.e., the slope of the fractional release vs time curve)

Swelling Controlled Devices. A completely different release behavior is observed for hydrophilic polymers when water sorption is followed by significant polymer swelling. Limiting Case II sorption occurs when water absorption is associated with a front advancing at a constant rate into the confines of a glassy polymer. A sharp boundary separates the unpenetrated core from the uniformly swollen shell (Figure 1, right). The polymer relaxation and swelling are determined by the osmotic stresses at the moving boundary, generated by the presence of the penetrant *(16-18)*. The polymer relaxation and swelling remain constant as long as a constant local concentration persists. The drug release is controlled quantitatively by the invasion of the swelling solvent and by the solute counter diffusion in the swollen polymer. Zero order release kinetics may be achieved for a polymer with a

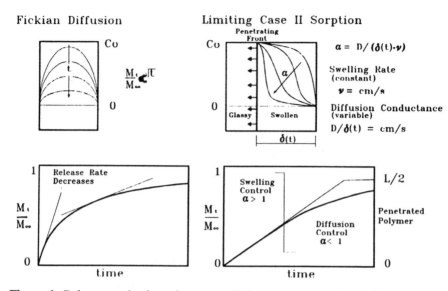

Figure 1. Release mechanisms from monolithic devices. (Reprinted with permission from ref. 27, Butterworth-Heinemann.)

constant penetration surface area, if the polymer swells at a constant rate, and if the counter diffusion of the solute molecules is rapid compared with the swelling rate.

Intermediate Cases. There are intermediate cases in which both the swelling and the diffusive control can be important during the drug release process. The advancing rate of the swelling front (n), and the diffusive conductance (the ratio between the solute diffusivity and the shell thickness at a given time $D/\delta(t)$) have been used to define the following dimensionless parameter α (2):

$$\alpha = D/[\ \delta(t) * n\ ]$$

which accounts for the relative contributions of the solute counter diffusion, and of the penetrant uptake rate to the overall rate of release.

At the early stages of swelling the diffusive conductance is high due to the small value of the swollen layer thickness ($\delta(t)$). The release is controlled by the polymer swelling rate (Figure 1). This condition provides α values greater than 1. Conversely, the diffusion of the solute molecules through the outer shell will change the observed release kinetics as the swollen layer becomes progressively thicker. In this case α values smaller than unity are observed. Diffusive control is reflected by the fractional release vs time curve as a progressive reduction of the release rate (Figure 1, right). For low values of α, the polymer rapidly swells and then the drug is depleted exclusively by a diffusive mechanism (2). As a consequence, a zero order release rate is expected when $\alpha(t) > 1$, and the penetration rate of the swelling agent is constant.

A theoretical concept regarding the penetrant uptake and the solute release was developed by Peppas *et al.* *(3,19)*, and by Davidson and Peppas *(20,21)* to characterize monolithic swellable systems for controlled drug release. It was shown that two dimensionless parameters should be used in order to predict the release behavior of these systems. The swelling interface number, **Sw**, ($1/\alpha$ in the present context) and the diffusional Deborah number, **De**, were introduced. The former represents the ratio of the penetrant uptake rate to the rate of solute diffusion. The latter represents the ratio of the characteristic swelling time of the polymer due to the presence of the swelling penetrant to the characteristic diffusion time of the penetrant into the polymer. Zero order release rates should be expected if both the solute diffusion through the swollen polymer layer is rapid compared to the penetrant uptake rate (**Sw** $<<$ 1), and the penetrant uptake is controlled by polymer relaxation (**De** = 1). As a consequence, the solute release kinetics would not be limited to α (or $1/$**Sw**). The value of **De** would also have to be taken into account.

Polymer Swelling and Dissolution Controlled Devices. Thermoplastic hydrophilic polymers are also water soluble. A sharp advancing front divides the unpenetrated core from the swollen and dissolving shell. Under stationary conditions, a surface layer (δ) of a constant thickness is formed by the swollen polymer and by a high concentration polymer solution *(22)*.

Once the hydrodynamic external conditions are defined, a stationary state is reached where the rate of penetration of the moving boundary (n) equals the rate of removal of the polymer at the external surface. The time interval until the quasi-stationary state is reached is called "swelling time" *(22)*.

Shown in Figure 2 is the typical polymer concentration in the surface layer of a dissolving polymer. If the dissolution occurs normally, the steady state surface layer consists of four different sublayers *(22)*: 1) liquid sublayer (adjacent to pure solvent), 2) gel sublayer, 3) solid swollen sublayer, and 4) infiltration sublayer (adjacent to polymer base into which the solvent has not yet migrated). If the test temperature is higher than the glass transition temperature of the polymer, then the surface layer consists only of the liquid and gel sublayers.

At steady state, the dissolution rate is constant and can be defined by either the velocity of the retracting front of the polymer, or the velocity of the front separating the pure penetrant and the liquid dissolving sublayer. Thus, the both fronts move synchronously.

The dissolution rate depends on hydrodynamic conditions, temperature, polymer molecular weight, and the degree of crystallinity. It has been found that in semicrystalline polymers, crystallization and dissolution behaviors occur in a similar way *(22)*. When plotted as a function of temperature, both the dissolution rate and crystallization rate show a maximum. Both decrease with increasing polymer molecular weight.

In the case of a dissolving polymer, a dimensionless parameter $\alpha(t)$ can still be defined as the ratio of the diffusive conductance $D/\delta(t)$ to the dissolution (or

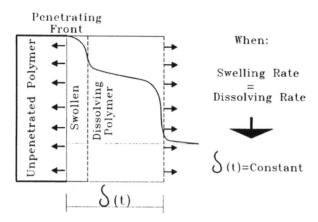

Figure 2. Polymer concentration profile for swelling and dissolving materials. (Reprinted with permission from ref. 27, Butterworth-Heinemann.)

penetration) rate. Consistent with the previous discussion on the swellable polymers, a zero order release rate can be expected only if $\alpha > 1$, and the dissolution rate is constant with time. Two different dissolution stages have been identified: 1) the initial time period (swelling time) and 2) the steady state conditions. During the initial transition, the thickness of the dissolving surface layer as well as the dissolution rate are not constant.

A time dependent diffusive conductance ($D/\delta(t)$) and a time dependent dissolution rate ($n(t)$) should be considered to evaluate the parameter $\alpha(t)$. Conversely, under steady state conditions, a constant diffusive conductance (constant gel layer thickness) and a constant dissolution rate are attained. As a consequence, at the first stage as α changes during the dissolution-release process the release rate can be a function of time. On the other hand, during the second stage, a time independent concentration profile develops in the external surface layer. A constant release rate is obtained which is determined by the penetrating front rate, or by the rate of the liquid sublayer-pure solvent boundary which is synchronous to the penetrating front. Similarly, for the delivery systems based on erodible polymer matrices, where the rates of diffusing and eroding fronts are identical, a zero order release kinetics is observed which is a result of the synchronous front velocities *(23)*.

Materials and Methods

Materials. Poly(ethylene oxide) (PEO) of average molecular weight 600,000 (Aldrich) and 4,000,000 (Aldrich) were used. ß-Hydroxyethyl-theophylline (etofylline) of analytical grade (purity 99.6%) was supplied by Sigma Chemical. The materials were used as received.

Tablet Preparation. The polymer and the powdered etofylline were dried under vacuum, mixed in the desired proportions, and then dissolved in chloroform. The solution was well stirred to ensure a homogeneous mixing of the components. Polymer films containing the drug were cast. After drying, several film layers were compression molded at 75 ° C to form sheets from which circular tablets (diameter 25 mm; 2.0 to 3.3 mm thick) were cut. Four different kinds of tablets were produced, all containing 10% etofylline. The four different polymer matrices used were: a) pure PEO (Mw 600,000), b) pure PEO (Mw 4,000,000) c) a blend of the two PEO's (1:1 by weight), d) a blend of the two PEO's (87 wt.% of 4,000,000 Mw and 13 wt.% 600,000 Mw; which is a 50% mol blend).

Calorimetric Analysis. To characterize PEO and PEO/etofylline mixtures, a DuPont Differential Scanning Calorimeter (DSC Instrument 910) operating under nitrogen at a heating rate of 10°C/min was used.

Gel Layer Thickness Measurement. The sample tablets were placed in thermostated distilled water at 37 ° C. The water was continuously stirred and aliquots were taken using a cathetometer. The degree of water swelling and penetration, and the gel layer evolution were optically measured at fixed times during the water conditioning.

Etofylline Permeation Tests through the Swollen Polymers. An apparatus equipped with a cell for measuring membrane liquid permeation was used for the permeability measurements (Absorption Simulator, Sartorius AG). Prior to the permeability measurements, the drug-free polymers were placed in the cell between two semipermeable cellulose acetate membranes and equilibrated with distilled water.

The water swelling at equilibrium in the cell was measured for the four different polymeric materials, and the data were used to calculate permeability values. The increase in the etofylline concentration in the downstream chamber of the permeability cell was monitored. The amount of the drug passing through the polymer matrix was evaluated as a function of time. Reference permeation tests were performed to determine the influence of the supporting cellulose acetate membranes on the drug permeation kinetics. The resistance to drug transport due to the supporting membranes was found to be negligible compared to permeation values for the swollen polymers. The drug concentration in the swollen polymer on the upstream side was determined from the permeation tests by calculating the partition coefficient for the water-swollen polymer and etofylline. The resulting concentration was approximately ten times less than the tablet drug loading. A drug concentration less than the drug loading is to be expected in the dissolving layer of the tablet due to dilution during swelling. Therefore, the diffusion constant obtained by the permeation tests was used to evaluate the dimensionless parameters introduced in the previous sections. The dependence of etofylline diffusivity through the surface layer with respect to composition was neglected.

Drug Release Kinetics Analysis. A dissolution apparatus (Erweka D.T.) operating at 50 rpm and at a constant temperature was used to evaluate the etofylline release kinetics from the polymer tablets. Each tablet was placed in a container filled with a known amount of distilled water, and the concentration of etofylline in the aqueous conditioning environment was measured as a function of time.

The etofylline concentrations in the aqueous solutions during both the permeability tests and the release tests were determined by UV spectroscopy using a Beckman DU-40 spectrometer operating at 262 nm. All the tests were carried out at 37° C.

Rheological and Viscoelastic Characterization. The rheological and viscoelastic characterization included small amplitude oscillatory shear flow tests and shear stress relaxation tests ("stress relaxation after sudden strain"). The first test consisted of measuring the unsteady response from a sample placed into a "cone and plate" geometry system, in which the plate was undergoing small amplitude sinusoidal oscillations, while the force required to keep the cone in position was continuously monitored.

In the second test, the sample is rapidly subjected to a small strain. The strain is then kept constant during the experiment, while the stress decrease is continuously monitored (stress divided by constant strain gives the shear relaxation module). The measurements were performed using a Bohlin VOR Rheometer with a cone-and-plate geometry. The tests were carried out on the polymer gel which formed the outer water-swollen and dissolving layers of the tablets placed in distilled water at 37° C.

We report in this presentation the value of tan δ as a function of shear oscillation frequency and the value of shear stress relaxation modules vs time in stress relaxation tests. Tan δ is defined *(24)* as a ratio of the loss shear modules, **G"**, related to the viscous character of the fluid, to the storage shear modules, **G'**, related to the elastic character.

Results and Discussion

Characterization of Drug Release Systems. Solvent penetration, polymer swelling and dissolution behavior, drug solubility in the host polymer, and diffusion in the swollen and gel-layer were examined in order to properly characterize the drug release from the swelling and dissolving systems discussed previously.

Differential Scanning Calorimetry. The DSC thermograms of etofylline, PEO, and of the tablets made of the four different PEO matrices containing 10% etofylline were obtained. The etofylline thermogram shows a well defined melting peak around 170° C, while those of the semicrystalline pure PEO's and the PEO blends showed melting peaks within the range of 63-65° C. The thermogram of the tablets made of the PEO blended with etofylline exhibited only a melting peak for the polymer matrix. The absence of an etofylline melting peak indicates a complete dissolution of the drug in the amorphous regions of the polymer. The crystalline fraction of the tablet matrices was about 65%, assuming that the crystallization enthalpy for PEO is -210 J/g *(25)*.

Polymer Swelling and Dissolution Properties. Reported in Figure 3 and 4 are the penetration kinetics for the tablets made of pure PEO with average molecular weights of 600,000 and 4,000,000, respectively, containing 10% etofylline. The 600,000 Mw PEO penetration curve as a function of time is almost linear (Figure 4). This indicates that after a very short initial transition the penetration front moves into the polymer matrix at a constant rate. The higher molecular weight

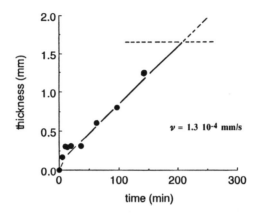

Figure 3. Penetration depth kinetics for water at 37°C in 600,000 Mw PEO tablets: the horizontal dotted line represents the time corresponding to the total penetration in a 3.3 mm thick tablet. (Reprinted with permission from ref. 27, Butterworth-Heinemann.)

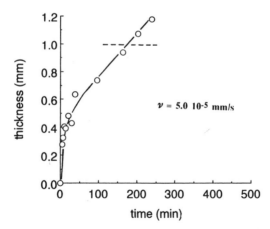

Figure 4. Penetration depth kinetics of water at 37°C for 4,000,000 Mw PEO tablets: the horizontal dotted line represents the time corresponding to the total penetration in a 2 mm thick tablet. (Reprinted with permission from ref. 27, Butterworth-Heinemann.)

PEO tablets (Figure 4) were characterized by an initially rapid water penetration rate. In the case of the lower molecular weight PEO, a sharp front moves at a constant rate through the tablet core after a significantly higher swelling time. The behavior of the PEO blends is similar to that of the pure 4,000,000 Mw PEO. The corresponding values for the steady state penetration rates, n, for the four kinds of polymer matrices are reported in Table 1. As expected, n decreases with increasing molecular weight. In fact, the PEO crystallization rate was reported to decrease with increasing molecular weight for Mw's exceeding $2x10^5$ - $8x10^5$. Similar behavior was observed in the case of dissolution rate *(26)*.

The evolution of the dissolving layer during the water conditioning should reflect the different dissolution characteristics of the investigated materials. When the dissolution rate equals the penetration rate, a surface layer of a constant thickness should be observed. The values of surface layer thickness as a function of time are compared in Figure 5. It is evident that a stationary thickness (about 1 mm) was achieved only for the lower Mw polymer. For the higher Mw PEO and for the two blends, the stationary thickness rapidly reached an initial value (1 mm) and then continuously increased (up to 5 mm) with further water sorption. Thus, for the given tablet thickness, a steady state dissolution rate (constant δ) was observed only for PEO with an average molecular weight of 600,000. The other three types of matrices did not reach the quasi-stationary dissolution stage. Based on this experimental evidence, a zero order release rate was expected for the tablets made of the Mw 600,000 PEO, while the other three matrices should have a more complex release behavior.

Drug Diffusivities in the Polymer Gel. Etofylline permeation tests in the water-swollen PEO were performed to evaluate the time lag. Shown in Figure 6 is the amount of the drug released in the downstream chamber at 37°C for the 4,000,000

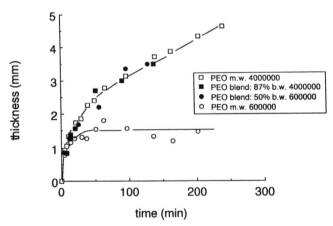

Figure 5. Gel layer development kinetics at 37°C for water penetrated PEO and PEO blends tablets. (Reprinted with permission from ref. 27, Butterworth-Heinemann.)

Mw PEO. The time lag is defined by the intercept with the time axis of the stationary part of the permeability curve. Assuming that the Fick's first and second laws for diffusive transport hold true, the time lag is equal to $L^2/6D$ where L is the equilibrium swelling thickness of the polymer sample, and D is the etofylline diffusivity. D can be readily evaluated from the time lag. The obtained values are reported in Table I.

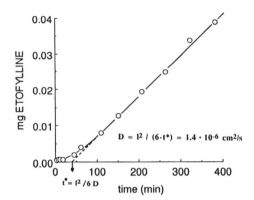

Figure 6. Etofylline permeation in distilled water swollen 4,000,000 Mw PEO. (Reprinted with permission from ref. 27, Butterworth-Heinemann.)

Drug Release Rate. Zero order drug release was expected for the tablets made of 600,000 Mw PEO. These tablets showed an almost constant release rate over the entire test period (Figure 7). Under the stationary conditions of swelling and dissolution, a dissolving surface layer of a constant thickness moved toward the

Table I. Swelling and Diffusion Parameters of PEO's

Polymer	$\cap$, cm/s	Gel Layer, cm	D, cm^2/s
PEO 600,000	13×10^{-6}	0.1	7.0×10^{-8}
PEO blend (50 wt.%)	7.0×10^{-6}	0.1 to 0.5	100×10^{-8}
PEO blend (50 wt.%)	7.6×10^{-6}	0.1 to 0.5	150×10^{-8}
PEO 4,000,000	5.0×10^{-6}	0.1 to 0.5	140×10^{-8}

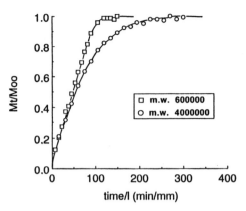

Figure 7. Release kinetics of etofylline at 37°C in distilled water from 600,000 and 4,000,000 Mw PEO tablets containing 10% by weight etofylline. (Reprinted with permission from ref. 27, Butterworth-Heinemann.)

inner part of the tablet. The drug distribution profile did not depend on time. The constant delivery rate was determined by the constant penetration (or dissolution) rate.

A more complex behavior was observed for the other three matrices studied. In the case of the tablets prepared with the 4,000,000 Mw polymer, the α value changed with time due to the growing thickness (δ) of the dissolving surface layer and to the decrease in the dissolving rate during the initial transition. From experimental data, $\alpha(t)$ ranged from greater than 1 to values smaller than 1 at the final stages. The development of drug diffusive resistance in the dissolving swollen layer became significant only when the thickness exceeds 2.5 - 3 mm. As a consequence, the drug release was governed by the dissolution rate during the initial stages of release, while at the final stages it was controlled by the solute diffusion through the swollen dissolving layer. The release kinetics from a tablet made of the 4,000,000 Mw PEO is reported in Figure 7. A period of an almost constant release rate was followed by a decreasing delivery rate.

The release tests for the tablets made of the two PEO's blends are reported in Figure 8. In both cases, the values of α ranged from greater than 1 to smaller than 1. Therefore, a behavior similar to that of the pure Mw 4,000,000 PEO should be expected. An almost constant release, which was very close to the behavior of the tablets made of pure 600,000 Mw PEO was observed in the case of the blend containing the lower amount of the higher Mw PEO (50% by weight). A release behavior similar, to that of the tablets made of the 4,000,000 Mw PEO, was observed only for the blend containing 87% by weight of the higher Mw PEO fraction. For the tablets made of the PEO's blends, α did not provide an interpretation of the data.

Rheological and Viscoelastic Characterization. The mucoadhesion of the polymeric devices depend on the viscoelastic properties of the polymer gel. The higher the elastic component of the polymer gel, the lower is the tendency of the tablet to flow from the site of adhesion. The viscoelastic properties change with the polymer molecular weight. The oscillatory data (Figure 9) showed that the elastic component of the polymer gel dramatically increased (decrease of tan δ) when the PEO's molecular weight rose from 600,000 to 4,000,000. The highest values of tan δ were obtained in the case of the 600,000 Mw polymer. The blends of the 600,000 and 4,000,000 Mw PEO's showed an increase in elastic behavior commensurate with the amount of the higher Mw PEO content. This indicates that the outer layer

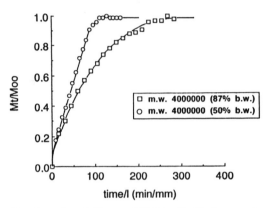

Figure 8. Release kinetics of etofylline at 37°C in distilled water from 50% by weight 4,000,000 Mw and 87% by weight 4,000,000 Mw PEO's blends tablets containing 10% by weight etofylline. (Reprinted with permission from ref. 27, Butterworth-Heinemann.)

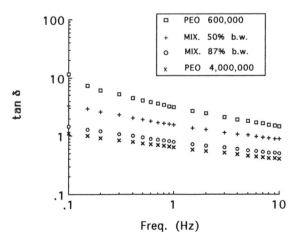

Figure 9. Tan δ vs oscillation frequency.

of the tablets tends to flow even if it is partially interpenetrated with the mucous layer. On the other hand the high elastic component of the 4,000,000 Mw polymer gel, coupled with the good mucoadhesive properties, provides good bioadhesion. Reported in Figure 10 is the polymer gel layer shear modules vs time (shear relaxation test) for different molecular weights. The modules increased with the amount of the 4,000,000 Mw fraction. A high value of the shear modules implies a good mechanical resistance to the stresses and deformations imposed by tissue movements.

The rheological and viscoelastic characterizations indicated that the optimal properties were observed in the case of the higher molecular weight polymer. Unfortunately, better release properties were obtained with the lower Mw PEO. A compromise between the viscoelastic and release properties was achieved by blending polymer fractions of different molecular weight.

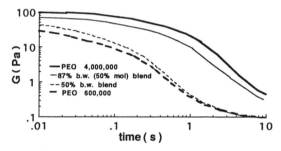

Figure 10. Gel layer shear modules vs time (Stress Relaxation Test).

Conclusion

It is possible to modify the release behavior of a monolithic drug delivery device by changing the molecular weight of a soluble polymeric matrix. The 600,000 Mw PEO matrix provided a nearly constant drug delivery rate due to synchronous penetrating and dissolving fronts. In the case of the 4,000,000 Mw PEO matrix, a non-constant release behavior was observed. The tablets made of a PEO blend which contained 87 wt.% of the high Mw fraction showed a behavior similar to that for the tablets made of the pure 4,000,000 Mw PEO. On the other hand, the application of the high Mw fractions with the proper bioadhesive and viscoelastic properties resulted in improved bioadhesion. The optimal results based on the release and the potential mucoadhesive properties were obtained for the tablets made of a blend (1:1 by weight) of the two PEO fractions. In this case, the tablets had a nearly constant release rate and good bioadhesive properties.

Literature Cited

1. Hopfenberg, H. B.; Hsu, K. C. *Polym. Eng. and Sci.* **1978**, 18, 1186.
2. Hopfenberg, H. B.; Apicella, A.; Saleeby, D. E. *J. of Membrane Sci.* **1981**, 8, 273.

3. Peppas, N. A.; Franson, M. N. *J. Polym. Sci., Polym. Phys. Ed.* **1983**, 21, 983.
4. Korsmeyer, R. W.; Peppas, N. A. *J. Controlled Release* **1984**, 1, 89.
5. Gaeta, S.; Apicella, A.; Hopfenberg, H. B. *J. Membrane Sci.* **1982**, 12, 195.
6. Good, W. R.; Mueller, K. F. *AIChE Symp. Ser.* **1981**, 77, 42.
7. Lee, P. I. *Polymer Comm.* **1983**, 24, 45.
8. Sefton, M. V.; Brown, L. R.; Langer, R. S. *J. Pharm. Sci.* **1984**, 73 (12), 1859.
9. Graham, N. B.; McNeill, M. E.; Zulfiqar, M. *Polymer Preprints* **1980**, 21, 104.
10. Graham, N. B.; Mc Neill, M. E.; Rashid, A. In *Advances in Drug Delivery Systems*, J. M. Anderson and S. W. Kim, Eds., Elsevier Science Publ., Amsterdam, 1986, 231-244.
11. Graham, N. B.; Mc Neill, M. E. *Biomaterials* **1984**, 5, 27.
12. Hunt, G.; Kearney, P.; Kellaway, I.W. In *Drug Delivery Systems*; P. Johnson and J. G. Lloyd-Jones, Eds.: Ellis Horwood Ldt Chichester and VHC Verlagsgesellschaft GmbH Weinheim, 1987, 180-199.
13. Sau-Hung; Leungand, S.; Robinson, J.R. *Polymer News* **1990**, 15, 333.
14. Chen, J.L.; Cyr, G.N. In *Adhesive Biological Systems*, R.S. Manly, Ed. Academic Press, New York and London, 1970.
15. Mikos, A. G.; Peppas, N. A. In *Bioadhesive Drug delivery Systems*, V. Lenaerts and R. Gurny Ed., CRC Press, Boca Raton Fl., 1990.
16. Sarti, G. C.; Apicella, A. *Polymer* **1980**, 21, 1013
17. Sarti, G. C.; Apicella, A.; DeNotariStefani, C. *J. Appl. Polym. Sci.* **1984**, 29, 4145.
18. Thomas, N. L.; Windle, A. H. *Polymer* **1981**, 22, 627.
19. Peppas, N. A.; Korsmeyer, R. W. In *Hydrogels in Medicine and Pharmacy*, Vol. III, N. A. Peppas, Ed., CRC Press, Boca Raton Fl., 1987.
20. Davidson III, G. W. R.; Peppas, N. A. *J. Controlled Release* **1986**, 3, 243.
21. Davidson III, G. W. R.; Peppas, N. A. *J. Controlled Release* **1986**, 3, 259.
22. Ueberreiter, K. In *Diffusion in Polymers*, J. Crank and G. S. Park, Eds., Academic Press, London and New York, 1968.
23. Lee, P. I. *J. Membrane Sci.* **1980**, 7, 255.
24. Bird, R. B.; Curtiss, C. F.; Armstrong, R. C.; Hassager, O. *Dynamics of Polymeric Liquids*, John Wiley and Sons, New York, 1987.
25. Price, C.; Evans, K. A.; Booth, C. *Polymer* **1975**, 16, 196.
26. Maclaine, J. Q. G.; Booth, C. *Polymer* **1975**, 16, 680.
27. Apicella, A.; Cappello, B.; Del Nobile, M. A.; La Rotonda, M. I.; Mensitieri, G.; Nicolais, L. *Biomaterials* **1993**, 14,83–89.

RECEIVED September 7, 1993

Chapter 10

Development of Micelle-Forming Polymeric Drug with Superior Anticancer Activity

M. Yokoyama[1,2], G. S. Kwon[1,2], T. Okano[1,2], Y. Sakurai[1,2], and K. Kataoka[2,3]

[1]Institute of Biomedical Engineering, Tokyo Women's Medical College, Japan
[2]Research Institute for Biosciences, International Center for Biomaterials Science and [3]Department of Materials Science and Technology, Science University of Tokyo, Yamazaki 2669, Noda-shi, Chiba 278, Japan

Polymeric micelles have been utilized as a drug carrier system which can efficiently incorporate hydrophobic drugs. This chapter presents a concept and strategy for using micelle-forming polymeric drug system with selective drug delivery. An example of a drug with enhanced *in vivo* anticancer activity is described.

Concept of Micelle-Forming Polymeric Drug

We have been utilizing polymer micelle architecture for the selective drug delivery to target the anticancer drug adriamycin *(1-6)*. Copolymers composed of hydrophobic and hydrophilic segments have the potential to form micellar structures in aqueous medium. Although block-copolymers of various types have the potential to form micellar structures, AB-type block-copolymer are the most appropriate candidates for designing the size, the aggregation number, and the stability of the micelles formed due to the simple molecular architecture.

Shown in Figure 1 is an AB-type block-copolymer composed of hydrophilic and hydrophobic segments which can form a micellar structure as the result of the amphiphilic character. The hydrophobic drug-containing segment forms hydrophobic core of the micelle, while the hydrophilic segment surrounds the core as a hydrated outer shell. Since most drugs have a hydrophobic character, these drugs are easily incorporated into the inner core segment by covalent bonding or noncovalent bonding such as hydrophobic interaction and ionic interaction. The advantages of a drug delivery system with polymeric micelles are summarized in Table I.

As described in Table I, the first advantage of micelle formation is the resultant long-term circulation in the blood as renal filtration is avoided due to the large size (polymeric micelles have much larger diameters than single polymeric chains). Even if the molecular weight of the constituting chains is lower than the critical molecular weight for renal filtration, these polymer chains can escape from

0097–6156/94/0545–0126$08.00/0

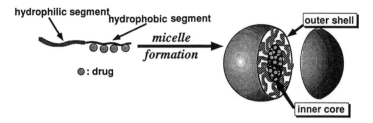

hydrophilic segment
hydrophobic segment
micelle formation
◎ : drug
outer shell
inner core

Figure 1. Architecture of block copolymer micelles.

Table I. Advantages of Micelle-Forming Polymeric Drugs

1.	Appropriate diameter for long half-life in blood
2.	No long-term accumulation
3.	High water solubility
4.	High structural stability
5.	Microreservoir in the hydrophobic core
6.	Separated functionality

renal excretion by forming the micellar structures with a larger diameter than the critical size for the renal filtration. As a result of the long-term circulation, the activity of the drug continues after one injection for a longer time.

On the other hand, since the polymeric micelles are formed by intermolecular noncovalent interactions in an equilibrium with a single polymer chain form, all polymer chains can be released as single polymer chains from the micelle structure with complete excretion by the renal route, provided that the polymer chains have a lower molecular weight than the critical values for the renal filtration. Therefore, polymeric micelles can be designed that are free of toxicities associated with long-term drug accumulation as the drug is circulating in the blood in the form of micelles.

For conventional polymeric carriers, the low water solubility of drug-polymer conjugates often causes synthesis problems *(7-9)* as well as injection problems *(10)*. Since most drugs have a hydrophobic character, the conjugation of the drugs with a polymer readily leads to precipitation due to the high localized concentration of hydrophobic drug molecules along the polymer chain. This is particularly the case when the hydrophilic functional groups (*e.g.*, amino and carboxyl groups) of the drug molecules are used for binding to the polymer and are changed into more hydrophobic groups (*e.g.*, amido group). Polymeric micelles with a core-shell structure can maintain their water solubility by inhibiting intermicellar aggregation of the hydrophobic cores, irrespective of high hydrophobicity of the inner cores. Thus, it is possible to utilize the hydrophobicity of the drug-binding segment, which causes precipitation in conventional polymeric drugs, to achieve a proper structure for drug targeting.

Block-copolymers composed of hydrophilic and hydrophobic segments are known to form micellar structures at much lower values of critical micelle concentrations (cmc) *(11)* than those for low molecular weight surfactants. The reason is that one block-copolymer molecule has more interaction sites for other polymers than low molecular weight surfactants. This can be utilized to form stable micelles under *in vivo* conditions for longer periods of time. The stable micelle formation makes it possible to control drug delivery by block-copolymers, on the basis of the micelle properties rather than on the properties of single polymer chains. As a result, a good correlation between *in vivo* and *in vitro* data can be obtained due to high *in vivo* reproducibility of micellar structures that are observed *in vitro*.

Drugs are incorporated into the core environment which is more hydrophobic than the exterior part of the micelles. The hydrophobic environment has a number of potentially useful properties. For example, inactivation of the drug molecules can be avoided by decreasing the contact with inactivating species in the aqueous (blood) phase (water and specific enzymes) *(12)*. Furthermore, a hydrophobic environment affects the release rate of the bound drug, in most cases decreasing the release rate. In other words, release rate is controlled both by stability of the micelles, and hydrophobicity of the micelle core and the chemical species used for binding the drug to the polymer backbone. In conventional polymeric drug,systems, the release rate is controlled mainly via the chemical bonds between the drug and the polymer.

In polymeric micelles, the functions required of drug carriers can be shared by structurally separate segments of block-copolymer. The outer shell is responsible for interactions with biocomponents, such as proteins and cells, which determine the pharmacokinetic behavior and biodistribution of the drug. Therefore, drug delivery *in vivo* is controlled by the outer shell segment, independent of the micelle inner core which expresses pharmacological activity. The heterogenous structure is more favorable for constructing a highly functionalized carrier system than the conventional polymeric carrier systems.

Polymeric Micelles as Drug Carriers

Except for our work, only a few studies have focused on the application of polymeric micelles as drug carriers. In 1984, Ringsdorf *et al. (13,14)* reported the concept of applying polymeric micelles to achieve sustained release. This was the first time that polymeric micelles for drug delivery systems were discussed. Micelle formation of this polymer-drug conjugate was suggested based on data for dye-solubilization and retardation in the release of a bound drug.

Kabanov et al. *(15)* reported an increase in the *in vivo* activity of a drug associated with the polymeric amphiphile, Pluoronic P-85 [poly(propylene oxide)-poly(ethylene oxide) block copolymer], as well as this amphiphile coupled to a specific antibody. They, however, did not distinguish the effect of micelle formation of the polymer-drug conjugate with respect to the efficient delivery to the target area compared to that of an increase in permeability of biological membranes due to the polymeric amphiphile. Also, they did not present evidence for micelle formation under physiological conditions. The authors *(13)* assumed the occurrence of successful and stable incorporation of a drug in the hydrophobic core of micelles formed by Pluoronic P-85. However, incorporation was not ensured under *in vivo* conditions because of the high critical micelle concentration (4.5%) for Pluoronic P-85.

We have obtained the first successful results with respect to a micelle-forming polymeric drug system, adriamycin-conjugated poly(ethylene glycol)-poly(aspartic acid) block-copolymer, (PEG-P[Asp(ADR)]). We found that (PEG-P[Asp(ADR)]) enhanced the *in vivo* activity of the drug, and confirmed the micellar structure under physiological conditions.

Enhanced Anticancer Activity of Micelle-Forming Polymeric Drug PEG-P[Asp(ADR)]

The chemical structure of the micelle-forming polymer-drug conjugate, adriamycin-conjugated poly(ethylene glycol)-poly(aspartic acid) block-copolymer, (PEG-P[Asp(ADR)]), is shown in Figure 2. By using PEG chain (Mw = 4,300) and poly(aspartic acid) chain (Mw = 1.900) with an ADR content of 30 mol.%, we found *(16)* that PEG-P[Asp(ADR)] produced higher anticancer activity against P 388 mouse leukemia with fewer side effects than free ADR.

Plots of the mean survival days for the treated mice versus the control mice (T/C), and body weight change for the treated mice 4 days after the drug injection against an ADR-HCl equivalent dose per kg of mouse body weight are shown in Figure 3. For ADR, a high value of T/C was obtained with a dose of 15 mg/kg. This dose, however, resulted in a marked decrease (15%) in the mice body weight.

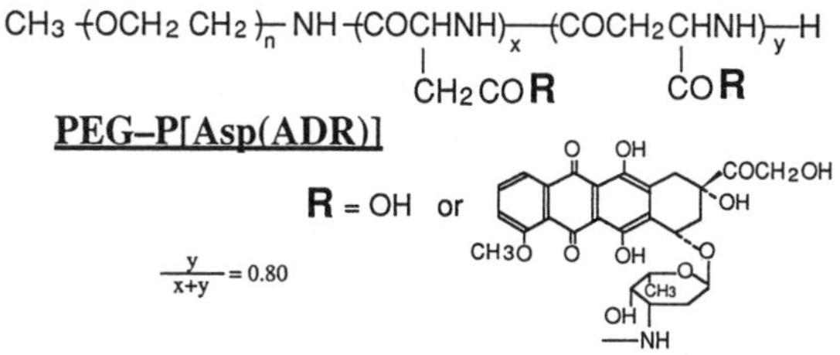

Figure 2. Chemical structure of adriamycin-conjugated poly(ethylene glycol)-poly(aspartic acid) block copolymer (PEG-P[Asp(ADR)]).

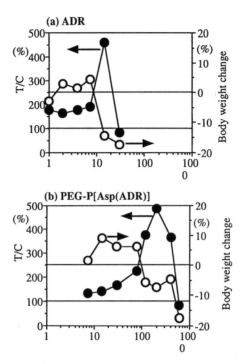

Figure 3. *In vivo* anticancer activity against P 388 murine leukemia. (a) ADR; (b) (PEG-P[Asp(ADR)]).

A dose of 30 mg/kg resulted in a shorter survival time compared to the control due to the acute toxicity. Thus, a high anticancer activity against P 388 leukemia was obtained only in the dose range that caused serious side effects (body weight loss). As for PEG-P[Asp(ADR)], the maximum value (>488%) was obtained at 200 mg/kg with a smaller decrease in body weight (7.4%) compared to the results for ADR at 15 mg/kg (15%). Furthermore, doses of 120 mg/kg and 400 mg/kg resulted in higher anticancer activity, with T/C values greater than 300%. This means that PEG-P[Asp(ADR)] has a wider effective dose range than ADR.

We also found that this micelle-forming drug-polymer conjugate possesses a superior antitumor activity against several solid tumors. The conjugates has a potential for the treatment of solid tumors which require high activity chemotherapy. The antitumor activity against murine colon adenocarcinoma 26 by PEG-P[Asp(ADR)] and by ADR is shown in Figure 4. Although ADR expressed inhibition of tumor growth, compared to the control mice, at its optimum dose of 10 mg/kg, the tumor volume never decreased from the initial size. At this dose of ADR, there were only long-term survivors (60 days) out of the five treated mice. In contrast, after treatment with PEG-P[Asp(ADR)] at the optimum dose, the tumor completely disappeared, and all the treated mice survived for 60 days. Though the targeting of anticancer drugs using drug delivery system has been

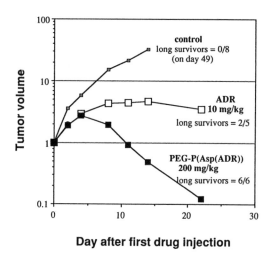

Day after first drug injection

Figure 4. *In vivo* anticancer activity against murine colon adenocarcinoma 26.

extensively studied, so far there is no successful example of greatly improved anticancer activity of a drug against solid tumor by conjugating the drug with a drug carrier.

ADR is one of the most active anticancer agents in present cancer chemotherapy. So why is the micelle-forming conjugate much more effective against C 26 tumor than ADR? The *in vivo* activity of the micelle-forming polymeric anticancer drug revealed a strong dependence on the composition, while the *in vitro* cytotoxic activity was found to be almost the same, regardless of the composition *(17)*. We think that selective delivery of the micelle-forming polymeric anti-cancer drug to the tumor is being achieved. Another question is why are micelles selectively delivered to the tumor, even without any specific targeting moieties such as antibodies? We speculate that micelles with a hydrated outer shell can be preferentially taken up and retained by the tumor in a manner similar to the enhanced permeation and retention of macromolecules reported by Maeda *et al. (18)*. To achieve this effect, it may be important to inhibit interactions of the hydrophobic core of the micelle with biocomponents by covering the core with the hydrated outer shell, which would provide a higher stability of the micelle structure in the bloodstream.

The pharmacokinetic behavior of ADR was drastically changed by incorporating into polymeric micelles. As shown in Figure 5, the concentration of PEG-P[Asp(ADR)] in the blood was kept significantly higher with a longer half-life compared to intact ADR. Mazzori *et al. (19)* reported that concentrations of ADR-bound poly(aspartic acid) in the blood were lower than those for intact ADR. The results indicate that the pharmacokinetic behavior of ADR in the plasma would not be changed by ADR binding to the homopolymer, probably because pharmaco-

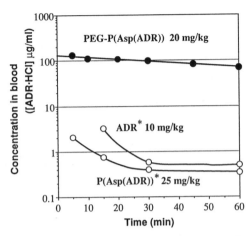

Figure 5.   Concentration in blood as a function of time after intravenous injection.

kinetic behavior was determined by the character of the bound ADR rather than by the polymeric carrier.

In contrast, the high concentrations and the long half-life of PEG-P[Asp(ADR)] in the blood indicate that ADR bound to the block-copolymer was stable in the circulation. Poly(ethylene oxide) is known as an inert synthetic polymer in living systems. It is utilized to decrease RES uptake *(20)* and to prolong a half-life in the blood *(21,22)*. Due to these properties, the outer shell of our micelles, which is composed of poly(ethylene oxide) chains, is thought to contribute to the stable circulation of PEG-P[Asp(ADR)] in blood by diminishing the interactions with biocomponents such as tissue, cells, and plasma proteins. This indicates that polymeric micelle architecture can be effectively utilized to control the *in vivo* delivery based on the characteristics of polymer rather than by the character of the bound drug.

Future Perspectives for Polymeric Micelles

For polymeric micelles, the pharmacokinetic behavior and biodistribution depend on the chemical character of the outer shell, as well as on the size and stability of the micelles (Figure 6). These three factors can be independent of the drug properties bound to the inner core. Therefore, drug delivery utilizing polymeric micelles can be independent of the pharmacologic agent carried.

Delivery of conventional polymeric drugs is affected by the drug bound to the polymer due to the drug interactions with biocomponents. Pharmacokinetically, various drugs can be utilized in polymeric micelle system by adjusting the quantity of the bound drug, and thus the hydrophobic-hydrophilic balance of the conjugate surface, required for micelle formation.

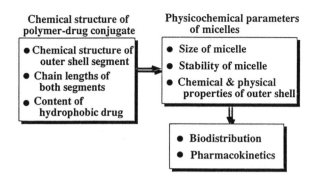

Figure 6. Control of *in vivo* delivery of polymeric micelles.

Polymeric micelle drug-carrying systems can be applied to numerous hydrophobic drugs. Furthermore, a hydrophobic drug can be physically entrapped in the hydrophobic core of the polymeric micelle without chemical bonding. At present, we are investigating the biologic activity of the polymeric micelle which physically entrap active drugs.

Literature Cited

1. Yokoyama, M., Inoue, S., Kataoka, K., Yui, N., Okano, T., Sakurai, Y. *Makromol. Chem.* **1989**, 190, 2041-2054.
2. Yokoyama, M., Miyauchi, M., Yamada, N., Okano, T., Sakurai, Y., Kataoka, K., Inoue, S. *Cancer Res.* **1990**, 50, 1693-1700.
3. Yokoyama, M., Miyauchi, M., Yamada, N., Okano, T., Sakurai, Y., Kataoka, K., Inoue, S. *J. Contr. Rel.* **1990**, 11, 269-278.
4. Yokoyama, M., Okano, T., Sakurai, Y., Ekimoto, H., Shibazaki, C., Kataoka, K. *Cancer Res.* **1991**, 51, 3229-3236.
5. Yokoyama, M., Kwon, G., Okano, T., Sakurai, Y., Seto, T., Kataoka, K. *Bioconjugate Chem.* **1992**, 3, 295-301.
6. Yokoyama, M., Kwon, G., Okano, T., Sakurai, Y., Ekimoto, H., Okamoto, K., Seto, T., Kataoka, K. *Drug Delivery* **1993**, 1, 11-19.
7. Hoes, C.J.T., Potman, W., van Heeswijk, W.A.R., Mud, J., de Grooth, B.G., Grave, J., Feijen, J. *J. Contr. Rel.* **1985**, 2, 205-213.
8. Duncan, R., Kopekova-Rejmanova, P., Strohalm, J., Hume, I., Cable, H.C., Pohl, J., Lloyd, B., Kopecek, J. *Br. J. Cancer* **1987**, 55, 165-174.
9. Endo, N., Umemoto, N., Kato, Y., Takeda, Y., Hara, T. *J. Immunol. Methods* **1987**, 104, 253-258.
10. Zunino, F., Pratesi, G., Micheloni, A. *J. Contr. Rel.* **1989**, 10, 65-73.
11. Wilhelm, M., Zhao, C.-L., Wang, Y., Xu, R., Winnik, M.A., Mura, J.-L., Riess, G., Croucher, M.D. *Macromol.* **1991**, 24, 1033.
12. Yokoyama, M., Miyauchi, M., Yamada, N., Okano, T., Sakurai, Y., Kataoka, K., Inoue, S. *Cancer Res.* **1990**, 50, 1693-1700.

13. Bader, H., Ringsdorf, H., Schmidt, B. *Angew. Chem.* **1984**, 123/124, 457.
14. Pratten, M.K., Lloyd, J.B., Hörpel, G., Ringsdorf, H. *Makromol. Chem.* **1985**,186, 725.
15. Kabanov, A.V., Chekhonin, V.P., Alakhov, V.Yu., Batrakova, E.V., Lebedev, A.S., Melik-Nubarov, N.S., Arzhakov, S.A., Levashov, A.V., Morozov, G.V., Severin, E.S., Kabanov, V.A. *FEBS Lett.* **1989**, 258, 343.
16. Yokoyama, M., Miyauchi, M., Yamada, N., Okano, T., Sakurai, Y., Kataoka, K., Inoue, S. *Cancer Res.* **1990**, 50, 1693-1700.
17. Yokoyama, M., Kwon, G., Okano, T., Sakurai, Y., Ekimoto, H., Okamoto, K., Seto, T., Kataoka, K. *Drug Delivery* **1993**, 1, 11-19.
18. Matsushima, Y., Maeda, H. *Cancer Res.* **1986**, 46, 6387-6392.
19. Mazzori, A., Gambetta, R.A., Trave, F., Zunino, F. *Cancer Drug Delivery* **1986**, 3, 163-172.
20. Lisi, P.J., Es, T. van, Abuchowski, A., Palczuk, N.C., Davis, F.F. *J. Appl. Biochem.* **1982**, 4, 19-33.
21. Abuchowski, A., MaCoy, J.R., Palczuk, N.C., Es, T. van, Davis, F.F. *J. Biol. Chem.* **1977**, 252, 3582-3586.
22. Katre, N.V., Knauf, M.J., Laird, W.J. *Proc. Natl. Acad. Sci. (U.S.A.)* **1987**, 84, 1487-1491.

RECEIVED June 18, 1993

Chapter 11

Optimization of Poly(maleic acid-*alt*-2-cyclohexyl-1,3-dioxepin-5-ene) for Anti-Human Immunodeficiency Virus Activity

Jian Ling Ding and Raphael M. Ottenbrite

Department of Chemistry, Virginia Commonwealth University,
Richmond, VA 23284–2006

A polymeric drug, poly(maleic acid-alt-2-cyclohexyl-1,3-dioxepin-5-ene) [poly(MA-CDA)], with molecular weights of 2,500 and 5,400 exhibited antiviral activity against HIV *in vitro*. These molecular weights were found to be too low for effective *in vivo* activity. To enhance the antiviral effect *in vivo*, the free radical copolymerization of maleic anhydride and 2-cyclohexyl-1,3-dioxepin-5-ene was optimized. The highest molecular weight was obtained by direct synthesis was 22,000 after solution fractionation. Poly(MA-CDA)'s with higher molecular weights were synthesized by polymer-polymer grafting. The original poly(MA-CDA) was modified with ethanolamine to form β-hydroxy amides along the polymer chain. The modified poly(MA-CDA) was then reacted with the original poly(MA-CDA) by means of their anhydride groups. The molecular weight of grafted products was controlled by varying the molar ratios of the modified polymer to the original polymer.

Polyanionic electrolytes are polymers that have negative charges along the molecular chain. The study of synthetic polycarboxylic acid anions in biological systems intensified in the early 1960's with the discovery that pyran (copolymer of divinyl ether and maleic anhydride) exhibited anticancer activity (1). Since then several synthetic polycarboxylic acid polymers have been evaluated and found to possess a variety of biological activities (2-6). These include stimulation of the reticuloendothelial system, modulation of humoral and/or cell mediated immune responses, and induction of resistance to microbial infections such as viruses, bacteria, protozoa, and fungi diseases.

The antiviral activity of polyanions is related to the molecular weight of the polymer and the polymer structure. Early work involving antiviral activity of natural and synthetic polyanions has been reviewed by Regelson (9) and Breinig et al. (8). Polyanions are cytotoxic against both DNA and RNA cytopathic viruses

0097–6156/94/0545–0135$08.00/0
© 1994 American Chemical Society

from several major virus groups with diverse characteristics. The ability of synthetic polyanions to afford protection to mammalian hosts against many viruses with varied properties is in contrast to the narrow antiviral spectrum of most conventional chemotherapeutic drugs. Most drugs are effective against a unique step in the replication of a particular virus or members of a particular virus group.

DeClercq and DeSome *(10)* reported that antiviral activity *in vivo* against vaccinia tail lesion formation increased with increasing molecular weight of poly(acrylic acid), whereas poly(methacrylic acid) was ineffective. Muck *et al. (11)* concluded that isotactic poly(acrylic acids) were considerably more active than atactic poly(acrylic acid) in providing protection against *picornavirus* infection *in vivo*, with optimum activity at Mw's ranging from 6,000 to 15,000. Morahan *et al. (12)* reported that one requirement for pyran to have potent antiviral activity against encephalomyocarditis virus is relatively high Mw (usually greater than 15,000). Our laboratory *(13)* compared the antiviral activity for pyran, poly(maleic acid-co-acrylic acid) [poly(MA-AA)], poly(maleic acid) [poly(MA)], and poly(acrylic acid-co-3,6-endoxo-1,2,3,6-tetrahydrophthalic acid) [poly(BCEP)]. Pyran and poly(MA-AA) produced approximately the same activity while poly(MA) and poly(BCEP) had very little effect. The low molecular weight fractions of these polyanions (<30,000) were ineffective. Only the higher molecular weight polymers were active. Synthetic polyanions can also provide prolonged protection against viral infection *(14,15)*. Compounds with similar molecular sizes, backbone structure and anionic character may differ considerably in their degree of antiviral activity *(16)*. It is important that the molecular weight of polymeric drug be less than 50,000 to allow for excretion from the host to minimize long term toxicity effects *(17)*.

This chapter is presented in three parts which relate to our research in this area. Part I deals with the anti-HIV activity of poly(maleic acid-alt-2-cyclohexyl-1,3-dioxepin-5-ene) [poly(MA-CDA)] and the problems associated with the polymer evaluation. Presented in Part II are the results of the optimization for the direct synthesis of poly(MA-CDA). Since the Mw required for a high *in vivo* activity with polycarboxylic acids is believed to be over 30,000 (12,13), Part III is a report on our attempts to increase the Mw by polymer-polymer grafting.

Part I. Anti-HIV Activity of Poly(MA-CDA)

Poly(maleic acid-alt-2-cyclohexyl-1,3-dioxepin-5-ene) [poly(MA-CDA)] (Figure 1) is a polycarboxylic acid polymer which elicits macrophage activation against tumor cells *in vivo (5,6,18)*. Poly(MA-CDA) exhibited greater macrophage activation than either pyran or *C. parvum* based on the *in vitro* morphological and [H^3]-thymidine evaluation against Lewis lung tumor cell. *In vivo* evaluation of poly(MA-CDA) demonstrated increased animal life spans which included survivors to Lewis lung carcinoma challenges *(18,19)*.

The antiviral activity of poly(MA-CDA) was evaluated recently. Poly(MA-CDA)s with different molecular weight, i.e. 2,500 (MA-CDA/2.5K) and 5,400 (MA-CDA/5.4K), were prepared and submitted to the U.S. Army Medical Research

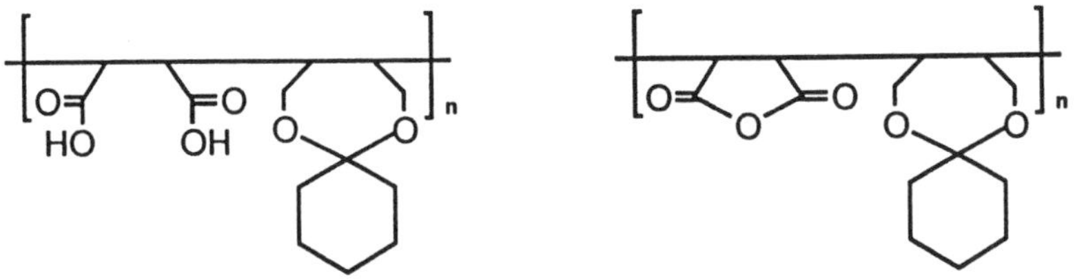

Poly(MA-CDA) Acid Form   Poly(MA-CDA) Anhydride Form

Figure 1. Structures of poly(MA-CDA).

Institute of Infectious Diseases. An MTT assay against human immunodeficiency virus (HIV) *in vitro*, the reverse transcriptase (HIV RT) screen *in vitro*, and *in vivo* test against Rauscher Leukemia Virus (RLV) were performed.

Methods

For the MTT assay, MA-CDA/2.5K and MA-CDA/5.4K were used to protect MT2 cell from HIV-3B. In an alternate cell line (CEM cells), MTT assay was repeated with MA-CDA/5.4K against another strain of HIV (HIV-CRF). In both cases, the cell viability was counted at different concentrations of the drug, with 5,2',3'-dideoxycytidine (DDC) and 3'-azido-2'3'-dideoxythymidine (AZT) as controls. The HIV reverse transcriptase screen was carried with ribosomal RNA as the template using both MA-CDA/2.5K and MA-CDA/5.4K in water or buffer at 37°C for one hour and the ID_{50} (50% inhibitory concentration) was determined.

The *in vivo* antiviral test involved three to four week old female BALB/c mice. MA-CDA/5.4K at three different doses was applied with AZT as control. Rauscher Leukemia Virus was administered i.p. on day 0, while the drugs were administered i.p. daily, beginning on day 1 through day 13. The mice were anesthetized, weighed and exsanguinated by axial cut. Spleens were removed and weighed. Sera was pooled, aliquoted and frozen at -70°C. UV-XC titers and RT levels were measured in serum samples.

Results and Discussion

The results of the MTT assays with MT2 cells using MA-CDA/2.5K and MA-CDA/5.4K against HIV-3B are shown in Figure 2. MA-CDA/2.5K and MA-CDA/5.4K exhibited antiviral effects at concentrations of 3.2 and 9.0 μg/mL when the CPE reductions were 10%. Shown in Figure 3 are the results of MTT assay with the CEM cell line using MA-CDA/5.4K against HIV-CRF. MA-CDA/5.4K provided 10% CPE reduction at concentration of 0.32 μg/mL at which it protected CEM cells from HIV-CRF. In both cell lines, MA-CDA/5.4K and MA-CDA/2.5K achieved 100% CPE reduction against different HIV viruses at a concentration of

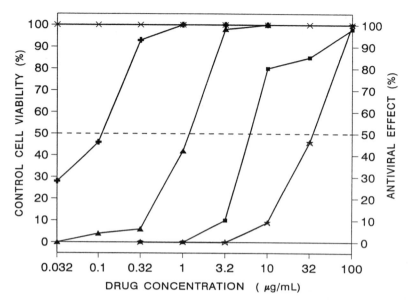

Figure 2. *In vitro* antiviral results of MTT assay with MT2 cells and HIV-3B virus. ×: drug's average cytotoxic effect (% cell viability) for all the drugs; Drug's antiviral effect (% reduction in viral CPE): + AZT, ▲ DDC, ■ MA-CDA/5.4K, ★ MA-CDA/2.5K.

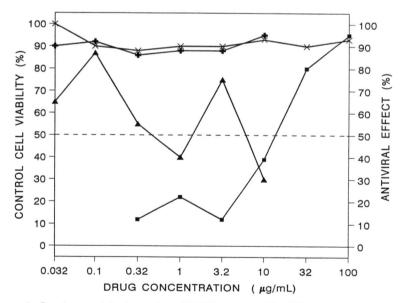

Figure 3. *In vitro* antiviral results of MTT assay with CEM cells and HIVCRF Virus. ×: drug's average cytotoxic effect (% cell viability) for all the drugs; Drug's antiviral effect (% reduction in viral CPE): + AZT, ▲ DDC, ■ MA-CDA/5.4K.

100 μg/mL. Both control drugs, AZT and DDC, provided 100% CPE reduction at lower concentrations. The general order of anti-HIV activity is AZT > DDC > MA-CDA/5.4K >MA-CDA/2.5K. The experimental data from the MTT assay are summarized in Table 1.

The results obtained from the reverse transcriptase screen are listed in Table 2 for MA-CDA/2.5K and MA-CDA/5.4K. Both poly(MA-CDA)s with a relatively low ID_{50} inhibited the virus replication *in vitro*. Thus, it appears that the mechanism of the anti-HIV action of poly(MA-CDA) may involve the transcription of RNA to DNA stage.

Table 1.   Poly(MA-CDA)'s Cell Line Assay Against HIV

Drug	Cell: MT2; Virus: HIV-3B			Cell: CEM; Virus: HIV-CRF		
	VD_{10} (μg/mL)	VD_{50} (μg/mL)	VD_{100} (μg/mL)	VD_{10} (μg/mL)	VD_{50} (μg/mL)	VD_{100} (μg/mL)
MA-CDA/2.5K	9.0	34	100	-	-	-
MA-CDA/5.4K	3.2	6.2	100	0.32	0.14	100
DDC	0.36	1.2	10	<0.032	0.44	-*
AZT	<0.032	0.11	1.0	NA	NA	NA

VD_{10} = concentration at which CPE reduction is 10%
VD_{50} = concentration at which CPE reduction is 50%
VD_{100} = concentration at which CPE reduction is 100%
NA = Not available; * = highly cytotoxic to cell as well.

Table 2. HIV Reverse Transcriptase (RT) Screen with Ribosomal RNA as the Template

Drug	Solvent	ID_{50} (μg/mL)	Effect on RT
MA-CDA/5.4K	H_2O	< 10	Inhibits
MA-CDA/5.4K	H_2O	< 10	Inhibits
MA-CDA/5.4K	H_2O	4	Inhibits
MA-CDA2.5K	Buffer	15	Inhibits
MA-CDA/2.5K	Buffer	2	Inhibits
MA-CDA/2.5K	Buffer	3	Inhibits

The results of poly(MA-CDA) *in vivo* test against Rauscher Leukemia Virus are listed in Table 3. It was found that the average animal weight decreased with increasing the dose of MA-CDA/5.4K. The spleen weight increased at all the dose levels, and the reverse transcriptase activities did not decrease after treatment of MA-CDA/5.4K. These data suggest that poly(MA-CDA) with molecular weight of 5,400 is not effective against RLV.

Poly(MA-CDA) with average Mw of 2,500 and 5,400 were active *in vitro* against HIV in both MTT assay and HIV reverse transcriptase screen. However, the *in vivo* test showed that poly(MA-CDA) with molecular weight of 5,400 was inactive against Rauscher Leukemia Virus. To achieve antiviral activity *in vivo*, Mw's of polyanions should be higher than 15,000, while Mw's higher than 30,000 may be needed for effective antiviral activity.

Part II. Direct Synthesis of Poly(maleic Acid-alt-2-cyclohexyl-1,3-dioxepin-5-ene)

Poly(MA-CDA) was first synthesized by Culbertson and Aulabaugh when they studied the copolymerization behavior of 4,7-dihydro-1,3-dioxepins and maleic anhydride *(20,21)*. However, a detailed copolymerization study of this monomer pair has not been reported. The average molecular weight of the MA-CDA (anhydride form) reported in literature was less than 8,000 *(20)*. The low molecular weight product may be due to the high steric hindrance associated with both monomers and the possible chain transfer at allylic methylene of CDA during the copolymerization. It is well documented that copolymerization of maleic anhydride and disubstituted olefins yield polymers with molecular weight lower than 10,000 due to steric hindrance imposed when both monomeric species are 1,2-disubstituted *(22,23)*. To synthesize high molecular weight poly(MA-CDA), the copolymerization behavior needs to be studied thoroughly.

Table 3. *In Vivo* Test of Poly(MA-CDA) against Rauscher Leukemia Virus

Drug Treatment	Avg AW[1] (% Change)	Avg SW[2] (% Change)	Avg RT Act[3] (% Change)	Survival (%)	Comment
MA-CDA/5.4K 1 mg/kg	-3.31	17.3	-11.3	90	Non-effective
MA-CDA/5.4K 3 mg/kg	-6.08	5.9	19.3	100	Non-effective
MA-CDA/5.4K 10 mg/kg	-9.39	20.6	91.2	90	Non-effective
AZT 0.3 mg/kg	-7.18	-8.4	-98.5	100	Effective

[1]Percent change of average animal weight (AW). [2]Percent change of average spleen weight (SW). [3]Percent change of average reverse transcriptase activity (RT Act).

We report here the optimization of direct free radical copolymerization of maleic anhydride and CDA and the isolation of different molecular weight poly(MA-CDA) samples by solution fractionation.

Experimental

Materials. 2-cyclohexyl-4,7-dihydro-1,3-dioxepin (CDA) was prepared according to literature procedures *(24)* and purified by vacuum distillation. Maleic anhydride (Aldrich) was purified by sublimation. The purity of all the monomers was checked by gas chromatography. AIBN (Polysciences) was purified from methanol. Polystyrene standards for gel permeation chromatography were purchased from Aldrich. All solvents were commercially available and purified before use by standard methods.

Optimization of Maleic Anhydride and CDA Copolymerization. A typical copolymerization process was carried out as follows. The reaction mixture was charged into a glass ampule and degassed three times by the freeze-thaw method. The ampule was sealed under vacuum and placed in a bath held at a specific temperature for specific periods of time. The polymer was precipitated by pouring the reaction mixture into anhydrous diethyl ether. The polymer was purified by reprecipitation from acetone in diethyl ether and dried in vacuum until a constant weight was achieved.

The copolymerization studies were performed by changing the ratio of monomers in the reaction mixture and evaluated below 10% of monomer conversion. At 60°C, the monomer molar ratios were regulated at 4:1, 3:1, 1:1, 1:3, 1:4, whereas the concentrations of the initiator remained unchanged. This study provided information regarding the copolymerization behavior and optimum monomer feed ratio for producing a higher molecular weight copolymer.

The copolymerizations were also carried out at 60°C with azobisisobutyronitrile (AIBN) as initiator while concentrations of the initiator or concentration of the monomers were varied. This provided the information regarding the optimum concentrations of initiator and monomers with respect to molecular weight.

Characterization. The polymer structure was characterized by ^{1}H-NMR and elemental analysis. The average molecular weight of polymers was determined by GPC-HPLC with a Waters μ-Styragel 10^3 Å column in THF at 30°C on a Perkin-Elmer 410 HPLC station. The GPC column was calibrated by using narrow polydispersity polystyrene standards.

Fractionation of MA-CDA. Poly(MA-CDA) with an average Mw of 11,000 and polydispersity of 3.1 was fractionated by solution precipitation. The poly(MA-CDA) was dissolved in tetrahydrofuran at a concentration of 1% by weight. Hexane was incrementally added as non-solvent. This process resulted in eight poly(MA-CDA) fractions at 30°C.

Results and Discussion

Copolymerization of MA and CDA. Shown in Figure 4 is the relationship between the monomer feed ratio of the starting comonomers and the average molecular weight of resultant polymer and the rate of polymerization. Neither MA or CDA homopolymerize under our experimental conditions. It is known that copolymerization of maleic anhydride with an electron donor, such as 4,7-dihydro-1,3-dioxepin derivatives, produces alternating copolymers (22,23). As illustrated in Figure 4, the highest rate of copolymerization was found at a monomer feed ratio of 1:1, while the highest Mw was obtained at 30% (mol) MA in the monomer feed. The copolymer composition was determined from ^{1}H-NMR and elemental analysis (will be reported elsewhere). These data support the fact that the poly(MA-CDA) is an alternating copolymer.

The effect of initiator concentration on Mw is shown in Figure 5. In general, Mw tends to increase with decreasing concentration of initiator. The Mw of poly(MA-CDA) was proportional to the monomer concentration (Figure 6). The highest molecular weight (11,700) was obtained in bulk with 20 mM of AIBN at 60°C. A poly(MA-CDA) with a Mw of 11,000 and a polydispersity of 3.1 was fractionated by a solution precipitation method. The highest Mw of the fractionated poly(MA-CDA) was 21,700 with a polydispersity index of 1.4 (Figure 7).

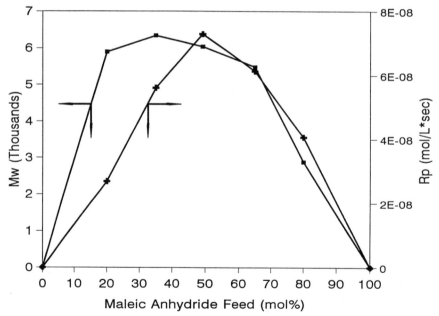

Figure 4. Effect of monomer feed ratio on molecular weight (■) and initial rate of polymerization (+) in the study of MA and CDA copolymerization in benzene at 60.0°C. [Comonomer] = 4.5 M; [AIBN] = 20 mM.

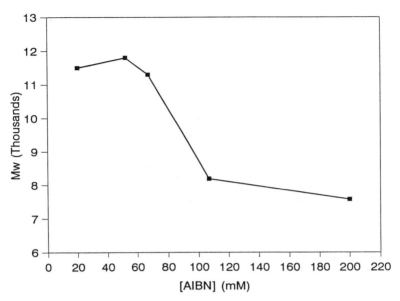

Figure 5. Effect of initiator concentration on Mw in the copolymerization of MA and CDA in benzene at 60.0°C. MA:CDA = 1:1 (mol); [Comonomer] = 6.4 M.

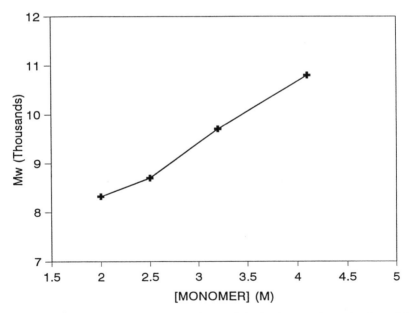

Figure 6. Effect of monomer concentration on Mw in the copolymerization of MA and CDA in benzene at 60.0°C. [AIBN] = 20 mM; MA:CDA = 1:1 (mol); The monomer concentration of 4.4 M is bulk condition.

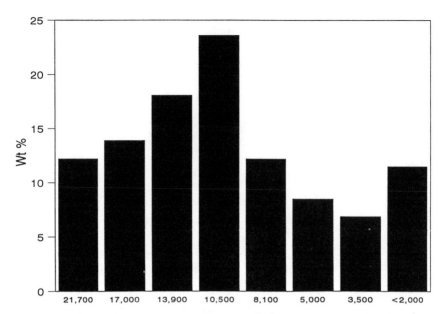

Figure 7. Solution fractionation of poly(MA-CDA). Original polymer: Mw = 11,000, Polydispersity index = 3.1. Resultant polymers: polydispersity index = 1.4-1.6.

In conclusion, direct free radical copolymerization of MA and CDA yielded poly(MA-CDA) with a Mw of 12,000. The highest Mw obtained by solution fractionation of poly(MA-CDA) was 22,000 with polydispersity index of 1.4. A method other than conventional free radical process may be needed to achieve molecular weight greater than 20,000.

Part III. Molecular Weight Augmentation through Polymer-Polymer Grafting

In order to achieve *in vivo* anti-HIV activity, the molecular weight of poly(MA-CDA) should be higher than 30,000 (12,13). However, direct copolymerization of MA and CDA yielded poly(MA-CDA) with a molecular weight of 12,000 and an isolated fraction with a Mw of 22,000. Therefore, a polymer-polymer grafting process was explored to synthesize poly(MA-CDA) with a molecular weight higher than 30,000.

Experimental

Materials. The parent poly(MA-CDA) with molecular weight of 3,300 (MA-CDA/3.3K) was synthesized by direct free radical copolymerization as reported above. Ethanolamine was purified by vacuum distillation. 4-Dimethylaminopyridine (4-DMAP) (Aldrich) was used as received. Polystyrene standards for gel permeation chromatography were purchased from Aldrich. All solvents were purified before use.

Polymer-Polymer Grafting. In a typical procedure, the polymer-polymer grafting was carried out as follows. To a flask equipped with a drying tube and a magnetic stirrer 293 mg (1.10 mmol) of MA-CDA/3.3K and 5.89 mL of N,N-dimethylformamide (DMF) were added and cooled to 0°C. To the polymer solution 133.0 mg (2.09 mmol) of ethanolamine in 1.5 mL of DMF were added dropwise. The reaction mixture was stirred at 0°C for 2 h and kept at 25°C for another 2 h. The modified polymer was isolated and dialyzed against distilled water at 0°C. The modified polymer was dissolved in 5.89 mL DMF again. To the modified polymer solution (1.18 mL, containing 0.181 meq. free hydroxyl groups) 45.9 mg (0.376 mmol) of 4-DMAP were added. To this solution 300 mg (1.13 mmol) MA-CDA/3.3K in 6.04 mL of DMF was added dropwise under stirring at 25°C. The reaction mixture was then warmed to 50°C for 4 h. The resultant polymer was precipitated in anhydrous diethyl ether, filtered, and dried under vacuum at elevated temperature.

Characterization. The polymer structure was characterized by ^{1}H-NMR and IR. The average molecular weight of polymers was determined by GPC-HPLC with a Waters μ-Styragel 10^3 Å column in THF at 30°C on a Perkin-Elmer 410 HPLC station. The GPC column was calibrated by using narrow polydispersity polystyrene standards. The intrinsic viscosity of polymers was measured with an Ubbelohde viscometer in 0.1 M potassium chloride DMF solution at 30°C.

Results and Discussion

The polymer-polymer grafting process is described in Scheme 1. Poly(MA-CDA) was first derivatized with ethanolamine to form repeat units of carboxylic acid-amide ethan-2-ol at 0°C. Polymer-polymer grafting was achieved through ester bond formation between the hydroxyl groups on the modified polymer chain and the anhydride groups of the original polymer chains at room temperature. The molecular weight of the products was controlled by varying the molar ratios of the modified polymer to the original polymer. Listed in Table 4 are the results for the polymer-polymer grafting study. The molecular weight of polymer-polymer grafted MA-CDA increased with increasing molars ratio of native poly(MA-CDA) to modified poly(MA-CDA). This observation was supported further by the viscosity data (Table 4). The intrinsic viscosity of the products increased with increasing amounts of anhydride MA-CDA in the reaction mixture. The resultant molecular weights were found to approximately match the expected molecular weights calculated from the sum of reactant ratios. The solubility information of starting polymers and products was reported in Table 5. The polymer-polymer grafted poly(MA-CDA)s are soluble in many organic solvents. The acid form of these products are soluble in aqueous solutions.

In conclusion, direct free radical copolymerization of MA and CDA can only yield poly(MA-CDA) with a Mw of about 12,000. By solution fractionation of this poly(MA-CDA), a Mw of 22,000 can be obtained, which is the highest Mw for this monomer pair, with a polydispersity of 1.4.

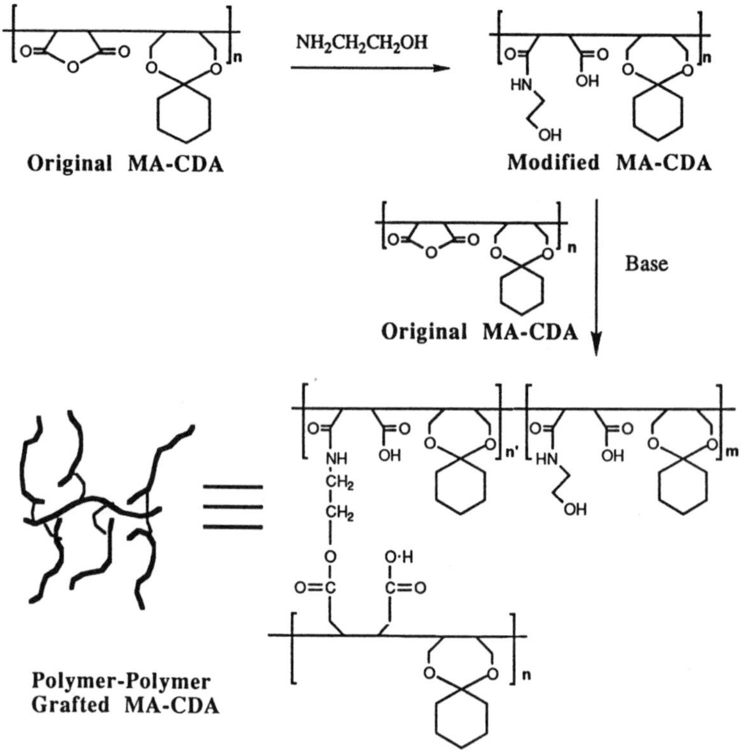

Scheme 1. Schematic description of polymer-polymer grafting.

Table 4. Experimental Results of Polymer-Polymer Grafting

Poly(MA-CDA) Ratios Original/Modified	η_{intr} (dL/g)	Mw expected	Mw[1] of Grafted Product
1:0	0.052	-	3,300
10:1	0.110	36,300	>30,000
7:1	0.091	26,400	29,500
6:1	0.080	23,100	26,700
3:1	0.068	13,200	N/A

[1]Peak average molecular weight obtained from GPC-HPLC

**Table 5. Solubilities of Polymer-Polymer Grafted Poly(MA-CDA)s.
(vs - very soluble; s - soluble; i - insoluble; s* soluble after the hydrolysis)**

Poly(MA-CDA) Ratios Original/Modified	THF	DMF	DMSO	Acetone	NaHCO$_3$ (0.4 M aq)
1:0	s	s	s	vs	s*
10:1	s	s	s	s	s*
7:1	s	s	s	i	s*
6:1	s	s	s	i	s*
3:1	i	s	s	i	s*

Polymer-polymer grafting provided poly(MA-CDA) with molecular weights higher than 30,000. The molecular weight of the final products can be controlled by varying the molar ratio of the original polymer to the modified polymer in the polymer-polymer grafting reaction. The samples of poly(MA-CDA) are being prepared for *in vivo* studies.

Acknowledgement. Authors wish to thank Dr. Jian-Zhong Yang for his valuable suggestions and technical assistance in the copolymerization study. We also thank the U.S. Army Medical Research Institute at Fredericksburg, MD, for the HIV analysis.

Literature Cited

1. Butler, G. B. In *Anionic Polymeric Drugs: Synthesis, Characterization and Biological Activity of Pyran Copolymers*, John Wiley & Sons, New York, p.49, 1980.
2. Ottenbrite, R. M. In *Anionic Polymer Drugs: Structure and Biological Activities of Some Anionic Polymers*, Donaruma, L. G., Ottenbrite, R. M., and Vogel, O., eds. John Wiley & Sons: New York; p.21, 1980.
3. Ottenbrite, R. M., Kuus, K., and Kaplan, A. M. *Polymer Preprints* **1983**, 24(1), 25.
4. Ottenbrite, R. M., Kuus, K., Kaplan, A. M. In *Polymer in Medicine*; Plenum Press: New York, pp.3-22, 1984.
5. Ottenbrite, R.M., Kuus, K., Kaplan, A.M. *J. Macromol. Sci.-Chem.* **1988**, A25, 873-893.
6. Ottenbrite, R. M., Takatsuka, R. *J. of Bioact. and Compat. Polymers* **1986**, 1, 46.
7. Regelson, W. *Adv. Chemther.* **1968**, 3, 303.
8. Breinig, M. C., Munson, A. T., and Morahan P. S. *Antiviral Activity of Synthetic Polyanions* in "Anionic Polymer Drugs", 1981.
9. Regelson, W., Morahan, P., and Kaplan, A. *Polyelectrolytes and Their Applications* D. Reidel Publishing Company, Dordrecht, 1975, p. 131

10. Declercq, E., and Desomer, P. *Appl. Microbiol.*, **1968**, 16, 1314
11. Muck, K. F., Rolly, H., and Burg, K. *Macromol. Chem.* **1977**, 178, 2773.
12. Morahan, P. S., Barnes, D. W., and Munson, A. E. *Cancer Treatment Rep.* **1978**, 62, 1797.
13. Ottenbrite, R. M. *Biological Activities of Polymers*, ACS Symposium Series, 186, p.205, 1982.
14. Claes, P., Billiau, A., DeClercq, E., Desmyter, J., Schonne, E., Vanderhaeghe, H., and Desomer, P. *J. Virol.* **1970**, 5, 313.
15. Schuller, B. B., Morahan, P.S., and Snodgrass, M. *Cancer Res.* **1975**, 35, 1915.
16. Merican, T. C., and Finkelstein, M. S. *Virol.* **1968**, 35, 363.
17. Duncan, R., Kopecek, J. In *Advances in Polymer Science: Soluble Synthetic Polymers as Potential Drug Carries*, New York and Berlin, 57, p.51, 1984.
18. Ottenbrite, R. M. *J. Macromol. Sci.-Chem.* **1985**, A/22(5-7), 819.
19. Ottenbrite, R.M., and Kaplan, A.M. *Ann. New York Acad. Sci.* **1985**, 446, 160.
20. Culbertson, B.M., Aulabaugh, A.E. *ACS Symp.* Ser. No. 195; pp.371-390, 1982.
21. Culbertson, B.M., Aulabaugh, A.E. *Polym. Preprints* **1981**, 22, 28.
22. Trivedi, B.C., Culbertson, B.M. *Maleic anhydride* Plenum Press: New York, 1982.
23. Cowie, J.M.G. (Ed) *Alternating Copolymers*, Plenum Press, New York, 1985.
24. Brannock, K.C., Lappin, G.R. *J. Org. Chem.* **1956**, *12*, 136.

RECEIVED August 23, 1993

Chapter 12

Enhancement of Cell Membrane Affinity and Biological Activity of Polyanionic Polymers

Yasuo Suda[1], Shoichi Kusumoto[1], Naoto Oku[2], and
Raphael M. Ottenbrite[3]

[1]Department of Chemistry, Faculty of Science, Osaka University,
Toyonaka, Osaka 560, Japan
[2]Department of Radiobiochemistry, School of Pharmaceutical Sciences,
University of Shizuoka, Shizuoka 422, Japan
[3]Department of Chemistry, Virginia Commonwealth University,
Richmond, VA 23284

To achieve a higher biological activity of polyanionic polymers for clinical applications, the cell membranes affinity of these polymers was enhanced by grafting hydrophobic groups onto the polymer. The membrane affinity was characterized *in vitro* using negatively charged liposomes, and *in vivo* using living rat intestine. The biological activities of the modified polyanionic polymers was also evaluated *in vitro* using cultured cell lines. In all cases, a higher biological activity was found with the modified polymer. The influence of the polyanionic polymers molecular weight on the biological response was also determined.

Several carboxylic acid polymers have been synthesized and examined for their biological response, such as cytotoxicity, antiviral activity, antitumor activity and immunomodulating activity *(1-7)*. Although biological results were significant, the polymer activity was considered insufficient for clinical applications. This limited activity may be attributed to low cellular interaction and uptake.

Different mechanisms for liposome and cellular uptake have been reported *(8,9)*. Sunamoto and Ottenbrite applied the technique of liposome encapsulation to improve the biological activity of polyanionic polymers *(10)*. The research was based on the concept that if the drug could be more easily incorporated into the target cell, a higher biological response might be expected. In fact, polyanionic polymers encapsulated in liposomes exhibited enhanced biological response. When the liposomes

0097–6156/94/0545–0149$08.00/0

POLYMERIC DRUGS AND DRUG ADMINISTRATION

were coated with the polysaccharide mannan, the activity increased 3 to 5 times *(5,10)*. Cell surface and mannan coated liposome recognition were thought to be involved in the cellular uptake mechanism. Liposome delivery presents an interesting technique, however, from the standpoint of clinical application, it may appear to be too complicated and expensive.

Shown in Figure 1 are some typical biological response modifiers from bacterial cell walls. These are negatively charged macromolecules *(11,12)* with highly hydrophobic regions within their structure. These hydrophobic regions appear to have a high affinity for cell membranes at an early stage of cellular interaction. It was rationalized that this hydrophobicity could also affect the biological activity of polyanionic polymers.

The most important stage in polyanionic polymer activity appears to be the initial interaction with the cell membrane. Once the polymer is in contact with the cell surface, the subsequent uptake of the polymer into the cell should occur readily. Normally, the interaction between cell membranes and polyanionic polymer is very weak because both the cell surface and polymers are negatively charged. To increase the membrane affinity, we have modified polyanionic polymers by grafting with

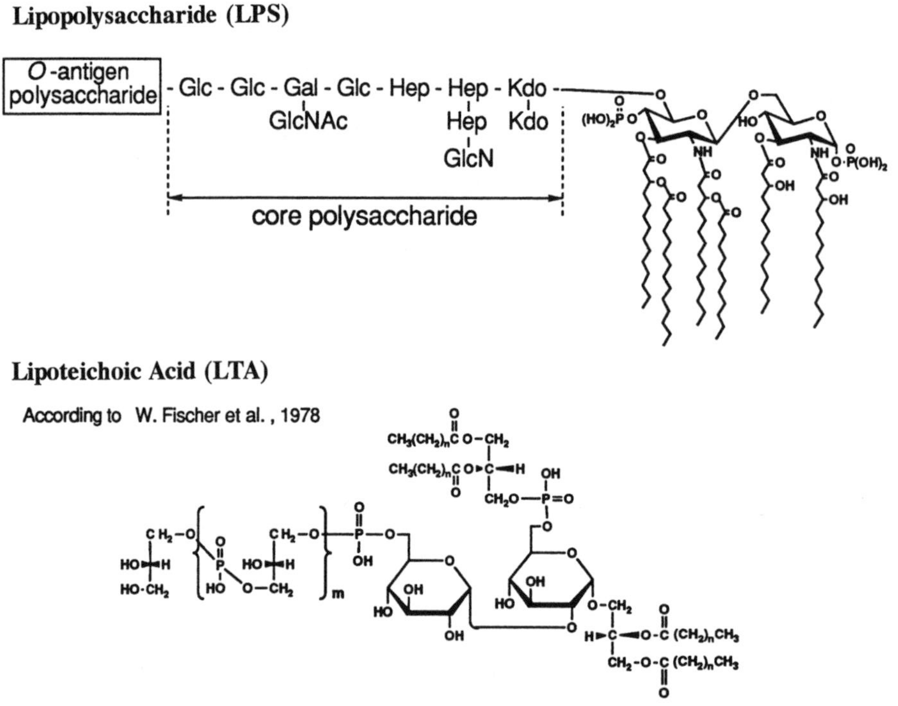

Lipopolysaccharide (LPS)

Lipoteichoic Acid (LTA)

According to W. Fischer et al., 1978

Figure 1. Estimated structures of the typical biological response modifier (BRM) from bacterial cell walls.

hydrophobic groups *(13)*. In this paper, we would like to report our recent work regarding hydrophobically modified polyanionic polymers.

Modification of Poly(maleic acid-alt-3,4-dihydroxyphenylprop-1-ene)

We studied hydrophobic groups seeking to improve membrane affinity while retaining water solubility for poly(maleic acid-alt-3,4-dihydroxyphenylprop-1-ene) (Figure 2) *(13)*. Hydrophobic amines were grafted onto the maleic anhydride residue of the polyanionic polymer with varying degrees of substitution (Table 1). Based on the solubility in water and partition coefficient data, the hexyl (H) and phenyl (A) groups appeared to be the more favorable hydrophobic functions to use.

Cell Membrane Affinity

a. Interaction with Liposomal Lipid Bilayer

The lipid bilayer of liposomes has been used as a cell membrane model to evaluate the affinity of the modified polymers *(14)*. The experimental principle is illustrated in Figure 3. An interaction of the polymer with the liposomal membranes usually causes a perturbation of the bilayer. This perturbation results in the leakage of calcein, a fluorescent compound, from the liposome. Calcein in high concentrations in the liposome is self-quenching, but the calcein leakage from liposomes produces a strong fluorescence. Therefore, the extent of the membrane perturbation can be estimated from the fluorescence intensity observed for the solution.

Shown in Figure 4 are the results with respect to the perturbation of the negatively charged liposomes by the modified polyanionic polymers *(13)*. The release of calcein was found to be dependent on the polymer concentration. The unmodified polymer, MA-DP, did not cause liposome perturbation suggesting that this polyanionic polymer is too hydrophilic to interact with cell membranes. On the other hand, the two modified polymers, MA-DP-A20 (phenyl grafted polymer with 20% loading) and MA-DP-H68 (hexyl grafted polymer with 68% loading) caused membrane perturbations even at low concentrations. The difference is 2 to 4 orders of magnitude compared with the unmodified polymer. This data indicates a stronger interaction between the grafted polymers and the liposomal bilayer. This suggests that a better interaction of the modified polymers with cell membranes may also be realized.

b. Interaction with Rat Intestinal Cells

The *in situ* loop method was applied to investigate interactions between the modified polymers and cell membranes (Figure 5) *(15)*. Compared with the unmodified polymer, the recovery of hexyl or phenyl grafted modified polymer from the rat small intestine was 10 to 20% less. This lower polymer recovery could be attributed to the adhesion of the polymer to the lipid bilayer of the intestinal epithelial cells and/or to absorption of the polymer by the cells.

Based on the data in Figures 4 and 5, the polyanionic polymer affinity for cell membranes appears to be enhanced by modification with hydrophobic groups.

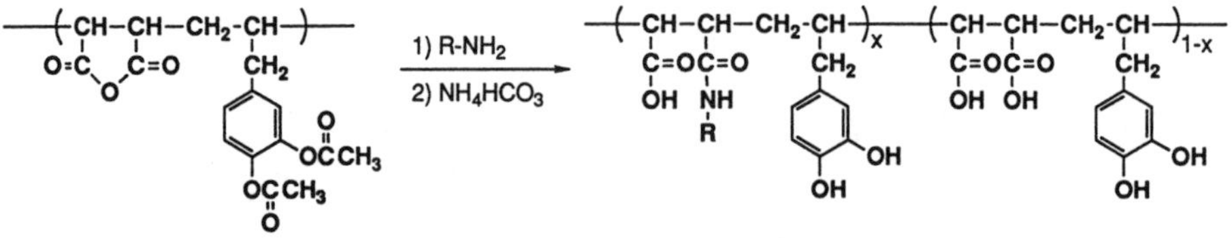

Figure 2. The modification of poly(maleic acid-alt-3,4-dihydroxyphenylprop-1-ene) by the grafting reaction.

Table 1. Modified poly(maleic acid-alt-3,4-dihydroxyphenylprop-1-ene)s and Their Partition Coefficients

Abbreviation	Grafting group	% Grafted	P.C.
MA-DP	none	0	0.002
MA-DP-A8	C_6H_5NH-	8	0.012
MA-DP-P23	$CH_3(CH_2)_2NH-$	23	0.026
MA-DP-B35	$CH_3(CH_2)_3NH-$	35	0.050
MA-DP-P100	$CH_3(CH_2)_2NH-$	100	0.058
MA-DP-O29	$CH_3(CH_2)_7NH-$	29	0.071
MA-DP-H30	$CH_3(CH_2)_5NH-$	30	0.088
MA-DP-D19	$CH_3(CH_2)_9NH-$	19	0.144
MA-DP-A13	C_6H_5NH-	13	0.243
MA-DP-B66	$CH_3(CH_2)_3NH-$	66	0.302
MA-DP-O59	$CH_3(CH_2)_7NH-$	59	0.309
MA-DP-H68	$CH_3(CH_2)_5NH-$	68	0.723
MA-DP-A20	C_6H_5NH-	20	0.868
MA-DP-B100	$CH_3(CH_2)_3NH-$	100	2.204
MA-DP-H100	$CH_3(CH_2)_5NH-$	100	n.d.
MA-DP-O100	$CH_3(CH_2)_7NH-$	100	n.d.
MA-DP-D74	$CH_3(CH_2)_9NH-$	74	n.d.

P.C.: Partition coefficient by n-octanol/phosphate buffer (pH 6.5).

n.d.: Not determined because of their insolubility into water.

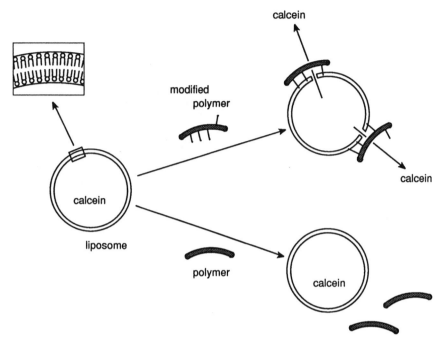

Figure 3. The principle for the evaluation of the interaction between modified or unmodified polyanionic polymer and liposomal membrane.

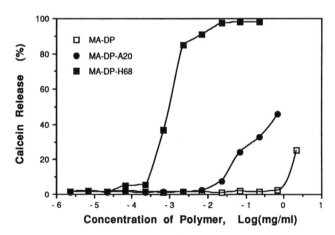

Figure 4. Effects of modified and unmodified poly(maleic acid-alt-3,4-dihydroxyphenylprop-1-ene) on the stability of the negatively charged liposome.

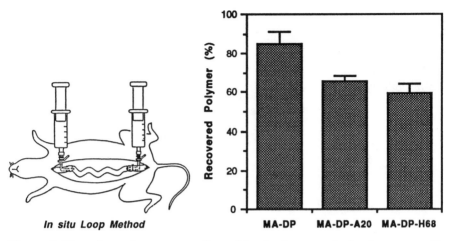

Figure 5. The schematic representation of in situ loop method using rat small intestine (left) and the percent recovery of the modified polymer from rat small intestine after 1 h (right).

Modification of poly(maleic acid-alt-2-cyclohexyl-1,3-dioxepin-5-ene)

Based on the modification results for poly(maleic acid-alt-3,4-dihydroxyphenylprop--1-ene), we applied the hydrophobic grafting technique to poly(maleic acid-alt-2-cyclohexyl-1,3-dioxepin-5-ene) *(16)*, which had previously been shown to have significant antitumor activity *(17)*. We prepared different molecular weight polymers (Figure 6) by changing concentration of the initiator (the molecular weights were estimated by using GPC-HPLC). The modification by grafting onto maleic anhydride residue was performed according the previous procedure *(13)*, with hexyl and phenyl groups as the hydrophobic groups.

Membrane Affinity

The *in vitro* affinity of the modified polymers for cell membranes was examined using a negatively charged liposome model *(14)*. Shown in Figure 7 are the results for a higher molecular weight polymer (Mw 20,000). In the case of the unmodified polymer (MA-CDA-20K), no calcein was released indicating poor interaction with liposomes. However, the phenyl modified (MA-CDA-20K-A20) and hexyl modified (MA-CDA-2-0K-H24) polymers did interact with the negatively charged liposomes. The results for the lower molecular weight polymers (Mw 3,400) are shown in Figure 8. The unmodified polyanionic polymer (MA-CDA-3.4K) exhibited little interaction, but the modified polymers (MA-CDA-3.4K-A18 and MA-CDA-3.4K-H27) interacted with liposomal membranes even at low concentrations. From the data in Figures 4, 5, 7, and 8, it may be concluded that the affinity of polyanionic polymers for cell membranes can be enhanced by simple hydrophobic group grafting.

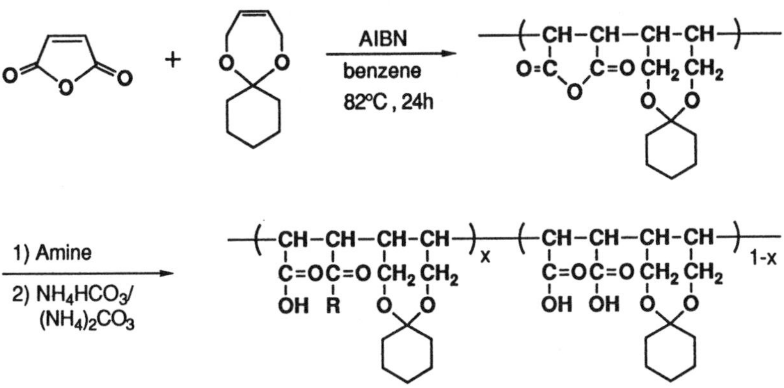

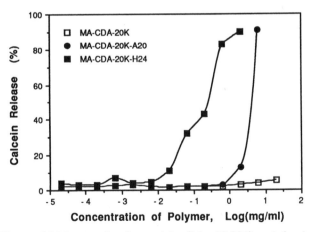

Abbreviation	MW of main chain	R	D.S. (x)
MA-CDA-20K	20000	HO-	—
MA-CDA-20K-H24	20000	CH₃(CH₂)₅NH-	0.24
MA-CDA-20K-A20	20000	C₆H₅NH-	0.20
MA-CDA-3.4K	3400	HO-	—
MA-CDA-3.4K-H27	3400	CH₃(CH₂)₅NH-	0.27
MA-CDA-3.4K-A18	3400	C₆H₅NH-	0.18

Figure 6. The preparation and the abbreviation of modified or unmodified poly(maleic acid-alt-2-cyclohexyl-1,3-dioxepin-5-ene).

Figure 7. Effects of higher molecular weight (Mw 20,000) poly(maleic acid-alt-2-cyclohexyl-1,3-dioxepin-5-ene)s on the stability of negatively charged liposome.

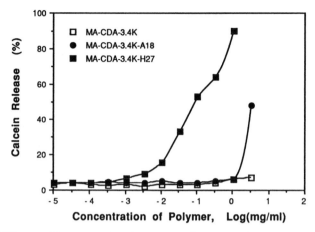

Figure 8. Effects of lower molecular weight (Mw 3,400) poly(maleic acid-alt-2-cyclohexyl-1,3-dioxepin-5-ene)s on the stability of negatively charged liposome.

Biological Activity

a. Superoxide Release

The biological activity of the polymers was evaluated *in vitro* by measuring the release of superoxide from DMSO-differentiated HL-60 cells *(18-20)*. The superoxide release was monitored using cytochrome C method *(21,22)*. As shown in Figure 9, when the differentiated cells were coincubated with a calcium ionophore A23187 (1.25 μg/mL, final concentration), the optical density at 550 nm increased. A23187 is a microbial product which acts as a perturbant of mammalian cell membranes. Since the superoxide increase was inhibited by the addition of superoxide dismutase (50 μg/mL, final concentration), it was concluded that this increase was due to the superoxide release from DMSO differentiated HL-60 cells. Similar results were observed for both modified and unmodified polymers. The effect of the higher molecular weight polymers on the release of superoxide from DMSO-differentiated HL-60 cells is shown in Figure 10. Each value is an average of at least three experiments and is expressed in relation to the A23187 control. The unmodified polymer displayed some stimulating activity only at high concentrations. In contrast, the modified polymer stimulated the superoxide release even at low concentrations, such as 0.027 mg/mL.

Shown in Figure 11 are the results for the lower molecular weight polymers. Compared to Figure 10, the modified low molecular weight polymers were found to have a much higher stimulating effect on the superoxide release. At high concentrations (0.34 mg/mL), the hexyl grafted polymer (MA-CDA-3.4K-H27) stimulated superoxide about 40 times greater than the control, and the phenyl grafted polymer

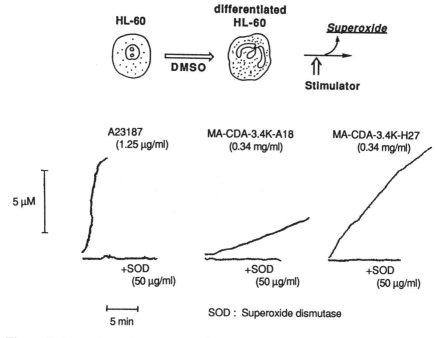

Figure 9. The schematic representation for the differentiation of HL-60 cells by the coincubation with dimethylsulfoxide [DMSO] (upper) and the assay of superoxide release in the DMSO-differentiated HL-60 cells by cytochrome C method (lower).

(MA-CDA-3.4K-A18) stimulation was 15 times greater. At lower concentrations, a distinct effect on the superoxide stimulation was seen in all of the modified polyanionic polymers.

Based on the information in Figures 10 and 11, hydrophobic group modification is an effective method for enhancing cellular response to polyanionic polymers which may relate to the improved cellular uptake of these substances.

b. Cytotoxicity

Qualitatively, the modified and unmodified polymers exhibited almost no cytotoxicity against DMSO-differentiated HL-60 cells during the superoxide release experiment (20-60 min). The long term effect of the polymers on cells was evaluated using cultured J774 macrophage cells (23).

Shown in Figure 12 are the effects of higher molecular weight polymers over a 2 day period on the growth of J774 cell which have a doubling time period of one day (16). It was assumed that the drug has cytotoxic effects when the cell growth is

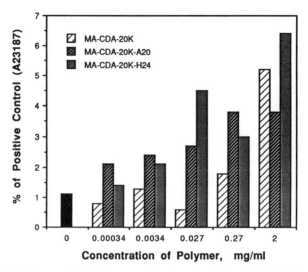

Figure 10. Effects of higher molecular weight poly(maleic acid-alt-2-cyclohexyl-1,3-dioxepin-5-ene)s (Mw 20,000) on the superoxide release in the DMSO-differentiated HL-60 cells. The data are the mean of at least three experimental results.

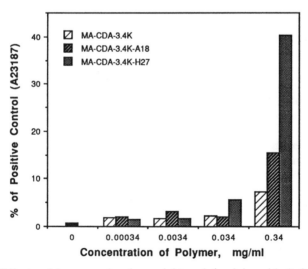

Figure 11. Effects of lower molecular weight poly(maleic acid-alt-2-cyclohexyl-1,3-dioxepin-5-ene)s (Mw 3,400) on the superoxide release in the DMSO-differentiated HL-60 cells. The data are the mean of at least three experimental results.

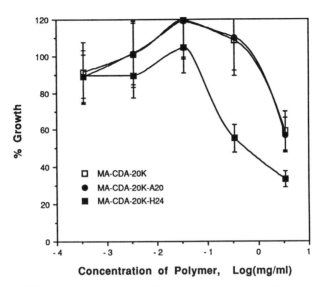

Concentration of Polymer, Log(mg/ml)

Figure 12. Effects of higher molecular weight poly(maleic acid-alt-2-cyclohexyl-1,3-dioxepin-5-ene)s (Mw 20,000) on the growth of J774 cells for two days.

lower than 25% compared to the control, and cytostatic effects between 25 and 50%. All three polymers showed no effect at concentrations less than 0.1 mg/mL. In the case of the hexyl modified polymer (MA-CDA-20K-H24), cytostatic effects were observed at 0.3 mg/mL, while the other two polymers showed cytostatic effects at levels greater than 1 mg/mL. The cytotoxic and cytostatic effects of the lower molecular weight polymers are shown in Figure 13. Compared to Figure 12, the cytotoxicity of all the lower molecular weight polymers seem to be greater. The hexyl modified polymer (MA-CDA-3.4K-H27) showed cytotoxicity at a concentration of 0.3 mg/mL. The higher cytotoxicity of hexyl modified polymers, MA-CDA-20K-H24 and MA-CDA-3.4K-H27, may be due to a greater cell membrane affinity.

It may also be possible that the perturbation of the membrane could directly stimulate the enzyme NADPH-oxidase located on the cell membrane that produces superoxide (24). To evaluate this possibility, an assay was performed using detergent (Triton X-100) instead of polymer. At 0.001% of Triton X-100, no stimulation of superoxide release from DMSO-differentiated HL-60 cells was observed. At 0.01% Triton X-100 exhibited a very strong cytotoxicity and the differentiated HL-60 cells were killed immediately. This data indicates that the membrane perturbation is involved in the cytotoxicity, but does not affect superoxide production.

A suggested mechanism for the expression of biological activity by modified polyanionic polymer is as follows: First, the polymer interacts with the cell membrane

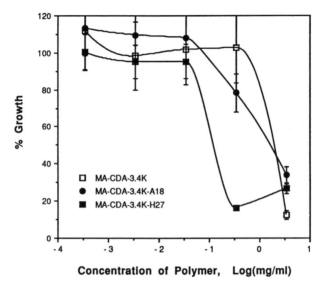

Concentration of Polymer, Log(mg/ml)

Figure 13. Effects of lower molecular weight poly(maleic acid-alt-2-cyclohexyl-1,3-dioxepin-5-ene)s (Mw 3,400) on the growth of J774 cells for two days.

due to the improved membrane affinity. Second, the polymer is incorporated into the cell and stimulates the signal transduction system inside the cell, where the interaction of polymer with protein kinase C (25) may take place. Finally, a signal is transduced to the membrane located NADPH-oxidase, which is activated to produce superoxide. The second stage seems to be affected by the molecular weight of the polyanionic polymer. Further evaluation of this hypothesis is planned.

Conclusion

In this paper, we have presented data that shows that polyanionic polymer affinity for cell membranes can be enhanced and controlled by the simple grafting of hydrophobic groups. We have performed two *in vitro* evaluations of these polyanionic polymers using cultured cell lines. The stimulation of superoxide release by the modified polyanionic polymers from the DMSO differentiated HL-60 cells was attributed to the enhanced affinity for cell membranes. The long term cytotoxicity on cultured J774 macrophage cells was only found at relatively high concentrations of polymers which may be related to the improved cellular affinity. In both tests, the molecular weight of the polymer affected the biological response. We are currently carrying out *in vivo* tests on the modified polymers reported. Further experiments are being performed to determine the effect of molecular weight on the biological activity.

Acknowledgement

The authors appreciate the help by H. Yamamoto, K. Shunto, A. Kawai, Y. Mori, and M. Miyashiro in carrying out these experiments. This work was supported in part by the Japan-US cooperation of the Japan Society for the Promotion of Sciences (JSPS) (Y. Suda) and the National Scientific Foundation, USA (NSF) (R. M. Ottenbrite).

Literature Cited

1. Merigancon, T.C. *Nature* **1967**, 214, 416.
2. Regelson, W., Munson, A., and Wooles, W. Interferon and Interferon Inducers, Int. Symp. Stand., London, 1969: *Symp. Series Immunobiol. Stand.*, Basel: Kager, S. 1978, 14, 227.
3. Donaruma, L.G., Ottenbrite, R.M., and Vogel, O., eds. *Anionic Polymeric Drugs.* NY: John Wiley and Sons, 1980.
4. Ottenbrite, R.M., and Butler, G.B., eds. *Anticancer and Interferon Agents.* NY: Marcel Decker, Inc., 1984.
5. Akashi, M., Iwasaki, H., Miyauchi, N., Sato, T., Sunamoto, J., and Takemoto, K. *J. Bioact. Compat. Polymers* **1989**, 4, 124.
6. Bey, P.S., Ottenbrite, R.M., and Mills, R.R. *J. Bioact. Compat. Polymers* **1987**, 2, 312.
7. Han, M.J., Choi, K.B., Chae, J.P., and Hahn, B.S. *J. Bioact. Compat. Polymers* **1990**, 5, 80.
8. Machy, P., and Leserman, L. *Liposomes in Cell Biology and Pharmacology*, John Libbey Eurotext, London and Paris, 1987.
9. Weinstein, J.N., and Leserman, L.D. *Pharmacol. Therapeut.* **1984**, 24, 207.
10. Sato, T., Kojima, K., Ihda, T., Sunamoto, J., and Ottenbrite, R.M. *J. Bioact. Compat. Polymers* **1986**, 1, 448.
11. Rietschel, E.T., Brade, L., Holst, O., Kulshin, V.A., Lindner, B., Moran, A.P., Schade, U.F., Zaehringer, U., and Brade, H. Endotoxin Research Series Vol.1, *Cellular and Molecular Aspects of Endotoxin Reactions*, Nowothy, A., Spitzer, J.J., Ziegler, E.J., eds., Excerpta Medica, Amsterdam-New York-Oxford, 1990, 15.
12. Fischer, W. *Handbook of Lipid Research* Vol.6, "Glycolipids, Phosphoglycolipids, and Sulfoglycolipids", Kates, M. ed., Plenum Press, New York and London, 1990, 123.
13. Suda, Y., Yamamoto, H., Sumi, M., Oku, N., Ito, F., Yamashita, S., Nadai, T., and Ottenbrite, R.M. *J. Bioact. Compat. Polymers* **1992**, 7, 15.
14. Oku, N., Yamaguchi, N., Yamaguchi, N., Shibamoto, S., Ito, F., and Nango, M. *J. Biochem.* **1986**, 100, 935.
15. Doluisio, J.T., Billups, N.F., Dittert, L.W., Sugita, E.T., and Swintosky, J.V. *J. Pharma. Sci.* **1969**, 58, 1196.
16. Suda, Y., Kusumoto, S., Oku, N., Yamamoto, H., Sumi, M., Ito, F., and Ottenbrite, R.M. *J. Bioact. Compat. Polymers* **1992**, 7, 275.
17. Kaplan, A.M., Kuus, K., and Ottenbrite, R. M. *Ann. New York Academy of Sciences* **1985**, 446, 169.

18. Collins, S.J., Gallo, R.C., and Gallagher, R.E. *Nature* **1977**, 270, 347.
19. Nath, J., Powledge, A., and Wright, D.G. *J. Biol. Chem.* **1989**, 264, 848.
20. Collins, S.J., Ruscetti, R.W., Gallagher, and R.E., Gallo, R.C. *Proc. Natl. Acad. Sci. U.S.A.* **1978**, 75, 2458.
21. Babior, B.M., Kipnes, R.S., and Curnutte, J.J. *J. Clin. Invest.* **1973**, 52, 945.
22. Meier, B., Radeke, H.H., Selle, S., Younes, M., Sies, H., Resch, K., and Habermehl, G.G. *Biochem. J.* 1989, 263, 539.
23. Raiph, P., Prichard, J., and Cohn, M. *J. Immunol.* **1975**, 114, 898.
24. Sumitomo, H., Takeshige, K., and Mizukami, S., *ENSHOU* **1985**, 5, 89.
25. Tsunawaki S., Kuratsuji, T. *FARUAW* **1990**, 26, 234.

RECEIVED October 12, 1993

Chapter 13

Accumulation of Poly(vinyl alcohol) at Inflammatory Site

Tetsuji Yamaoka, Yasuhiko Tabata, and Yoshito Ikada[1]

Research Center for Biomedical Engineering, Kyoto University, 53 Kawahara-cho, Shogoin, Sakyo-ku, Kyoto 606, Japan

The body distribution of a synthetic water-soluble polymer, poly(vinyl alcohol) (PVA), was studied in mice. The inflammation in mice was induced by a carrageenan injection. After the intramuscular injection of carrageenan into the mouse femora, ^{125}I-labelled PVA of different molecular weights was administered intravenously. The concentration of PVA accumulated at the inflammatory site was compared with that at the normal site. It was found that PVA tended to accumulate at the inflammatory site at higher concentrations than at the normal site. The PVA accumulation depended on its molecular weight. The maximum accumulation was observed for the Mw 200,000 PVA. The pharmacokinetic studies demonstrated that the accumulation rate of PVA at the inflammatory site was higher than at the normal site and decreased with an increase in the molecular weight. On the other hand, the higher Mw PVA was retained in the blood circulation for a longer period of time. The balance of the two factors, the PVA retention in the blood and the rate of PVA accumulation at the inflammatory site, affects the profile of the PVA accumulation at the inflammatory site, with the maximum accumulation for the Mw 200,000 PVA.

The chemical conjugation of drugs to biological and synthetic water-soluble macromolecules can modify their body distribution, increase therapeutic efficacy, and reduce adverse side effects. The fate of the drug conjugates in the body after administration depends on their physical properties, such as size and electric charge, and biological properties (1,2). We have investigated the effect of

---

[1]Corresponding author

0097–6156/94/0545–0163$08.00/0

molecular weight of various water-soluble polymers on half-life in the blood circulation and on tissue-distribution (3). It was found that the vascular permeability of polymers depended on their molecular weight, with a drastic change around Mw 30,000.

It has been reported (4) that the leakage of proteins and lipids from the blood vessels into the interstitial space is higher in the inflammatory tissues due to the enhanced permeability of the vascular walls. It has been further demonstrated (4,5) that inflammation affected the body distribution of polystyrene microspheres, as well as serum albumin, resulting in their higher accumulation at the inflammation site compared to the normal sites. However, the effect of physicochemical properties of synthetic water-soluble polymers on their behavior in the body with inflammatory processes was not studied.

The present work was undertaken to study the biodistribution of radiolabelled poly(vinyl alcohol) (PVA) in mice with carrageenan-induced inflammation. The effect of the molecular weight of PVA on its accumulation at the inflammatory sites was investigated.

Materials and Methods

Reagents. PVA was kindly supplied by Unitika Kasei Ltd., Osaka, Japan. The average molecular weights of the PVA samples, determined by gel filtration chromatography (Tosoh Co., Ltd., Tokyo, Japan) were 14,000, 68,000, 125,000, 196,000, and 434,000, with poly(ethylene glycol) used as a standard. The degree of saponification exceeded 98% for all PVA's. 1,1'-Carbonyl diimidazole (CDI) and sodium pyrosulfite (SMS) were purchased from Nakalai Tesque, Kyoto, Japan. Tyramine was obtained from Wako Pure Chemical Industries, Ltd., Osaka, Japan. $Na^{125}I$ solution and anion-exchange resin Dowex were purchased from NEN Research Products, Dupont, Wilmington, DE, and Dow Chemicals Co., Ltd., Midland, MI, respectively. The chemicals were of guaranteed grade and used without further purification.

Labeling PVA with [125]**I.** The radio iodination of PVA was carried out according to the conventional chloramine-T method (6). First, tyramine residues were introduced into hydroxyl groups of the PVA by use of CDI (7). Dimethyl sulfoxide (DMSO) (2 mL) containing 7.4 mg CDI was added to 50 mL of 0.4 wt.% PVA solution in DMSO. The mixture was stirred for 30 min at 25°C to activate the hydroxyl groups of the PVA. The reaction mixture was dialyzed against distilled water for two days, followed by the dialysis against 10 mM sodium borate buffer of pH 8.5. Then, 62.3 mg of tyramine were added to the activated PVA in the borate buffer, and the mixture was stirred at 25°C for 48 h. The resulting solution was lyophilized after the dialysis against distilled water for 2 days to obtain the tyramine-bound PVA. The UV measurements of tyramine absorbance at 280 nm demonstrated that the introduction of tyramine to the PVA molecule was about 0.15 mol.%, based on the number of hydroxyl groups in the PVA, regardless of its molecular weight. A solution of $Na^{125}I$ (2 μL) was added to 150 μL of 5 mg/mL tyramine-bound PVA in 0.5 M potassium phosphate buffer (KPB, pH 7.5). Chloramine-T (100 μL, 0.2 mg/mL) in 0.05 M KPB (pH 7.2) was added to the solution. After agitating for 2 min, 100 μL phosphate-buffered saline solution (PBS, pH 7.4), containing 0.4 mg SMS, was added to stop the iodination reaction. The resulting mixture was passed through the column of Dowex to remove the uncoupled free ^{125}I molecules from the ^{125}I-labelled PVA.

Measurement of PVA Concentration in Plasma and at the Inflammatory Site. BALB/cCrSlc mice (8-12 weeks old, Shizuoka Animal Facility Center, Shizuoka, Japan) received intravenous injection of 100 μL of 1% ^{125}I-labelled PVA solution in PBS. The blood samples (50 μL or less) were taken from the retroorbital plexus at different time intervals. The radioactivity remaining in the whole blood was determined from the radioactivity of the blood sample and the total volume of the mouse blood. The percent radioactivity was evaluated as the distribution volume of ^{125}I-labelled mouse serum albumin after intravenous administration.

The inflammation was induced in the mice by injecting carrageenan according to the method reported by Davis *et al.* *(5).* Carrageenan (100 μL of 1.5%) in saline solution was intramuscularly injected into the left femora, and the PBS containing ^{125}I-labelled PVA was administered through the tail vein three days after the carrageenan injection. At different time intervals after the intravenous administration of ^{125}I-labelled PVA, the femoral muscles were excised and washed twice with PBS. The excised tissues were weighed. The radioactivity in the inflammatory left femoral muscle, as well as in the normal right femoral muscle, was measured to estimate the concentration of PVA accumulated at both sites.

Pharmacokinetic Analysis. The accumulation rate of the PVA at the inflammatory and at the normal sites were calculated using the method of Takakura *et al.* *(8),* based on the one compartment model shown in Figure 1. The change of radio activity in tissue can be described by the following equation 1:

$$\frac{dA_{(t)}}{dt} = R_{acc} \times C_{(t)} - R_{eli} \times A_{(t)} \qquad (1)$$

where $A_{(t)}$ (% of dose/g) and $C_{(t)}$ (% of dose/mL) are the ratio of the remaining radioactivity per gram of the tissue, or per mL of the plasma, to the total radioactivity injected at time t, respectively. R_{acc} (mL/h/g) is the rate index for accumulation from the plasma to the tissue, and R_{eli} (1/h) is the rate constant for elimination from the tissue. Immediately after the ^{125}I-PVA injection, the second term in equation 1 is negligible since $A_{(t)}$ is sufficiently small compared to $C_{(t)}$. Ignoring the elimination term, equation 1 integrates to the following expression:

$$R_{acc} = \frac{A_{(t_1)}}{\int_0^{t_1} C_{(t)}\, dt} = \frac{A_{(t_1)}}{AUC_{0-t_1}} \qquad (2)$$

According to equation 2, the accumulation rate index can be calculated from the tissue radioactivity at time t_1 and from the area under the plasma concentration-time curve up to time t_1 (AUCt_1). The AUC, which is the measure of the concentration of the PVA passing through a site, was calculated by integrating the right part of the equation 3 which describes the time course of the radioactivity profile after the intravenous injection of ^{125}I-labelled PVA.

$$C_{(t)} = A \times e^{-\alpha t} + B \times e^{-\beta t} \qquad (3)$$

The parameters A, B, α, and ß in equation 3 were determined from the PVA concentration by curve fitting using the nonlinear least-squares program MULTI (8).

Results

PVA Concentration in Blood. Shown in Figure 2 is the effect of the PVA molecular weight on the PVA concentration in blood after an intravenous injection to normal mice. Apparently, the higher Mw PVA was retained in the blood circulation for a longer period of time. The profile of PVA plasma concentration was analyzed using equation 3 to obtain an AUC value for each PVA (Figure 6).

PVA Accumulation at the Inflammatory and Normal Sites. The concentration of PVA in the femoral muscles of the normal non-injected mice was similar to that for the right femoral muscle of the mice which had received a carrageenan injection in the left femora (data not shown). Thus, carrageenan injection had no effect on the PVA distribution at the normal, non-injected site.

Shown in Figure 3 is the effect of the time interval between the PVA and the carrageenan injections on the PVA accumulation at the inflammatory and normal sites. The PVA (Mw 196,000) was i.v. administered after the carrageenan injection. The concentration of the accumulated PVA was estimated by measuring the PVA radioactivity 24 h after its injection. The PVA concentrations at the

Figure 1. One compartment model of polymer accumulation.

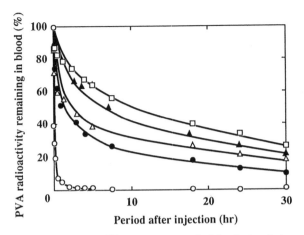

Figure 2. Decrement patterns of ^{125}I-labelled PVA in blood circulation with time after i.v. injection. PVA molecular weights were (o) 14,000; (●) 68,000; (▵) 125,000; (▴) 196,000, and (□) 434,000.

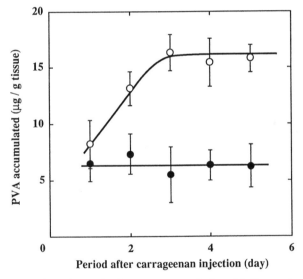

Figure 3. Accumulation of PVA in the carrageenan-induced inflammatory sites (o) and the normal sites (●) (Mw 196,000).

inflammatory site were higher than those at the normal site (Figure 3), irrespective of the period after the carrageenan injection. The PVA concentration at the inflammatory site increased with time after the carrageenan injection up to 3 days, and then it levelled off. At the normal site, the PVA concentration was constant during the test period. Thus, the difference in the PVA concentrations between the two sites increased with time after the carrageenan injection. This indicates that it takes a few days for the inflammatory vascular walls to acquire a significantly higher permeability to PVA. In the following experiments, the PVA was always injected 3 days after the carrageenan injection.

Effect of PVA Molecular Weight. The accumulation of PVA with different molecular weights was compared for the inflammatory and the normal sites (Figure 4). No significant difference in the PVA accumulation at the two sites was observed 3 h after the PVA i.v. injection, irrespective of the molecular weight. The PVA concentrations at the inflammatory site were significantly higher than at the normal site 24 h after injection (Figure 5). The accumulation of PVA depended on the molecular weight, with the maximum difference in the PVA concentrations for the Mw 196,000 PVA, irrespective of the period after the PVA injection.

PVA Accumulation at Severely Inflammatory Site. After a single carrageenan injection, only slight inflammation was observed at the injection site. To induce more severe inflammation, three carrageenan injections were given. The results (Table 1) indicate that the PVA accumulation at the inflammatory site was enhanced by the number of carrageenan injections. This may be explained by the increased vascular permeability caused by severe inflammation.

Discussion

Although not as remarkable as serum albumin *(4)*, PVA accumulated at the

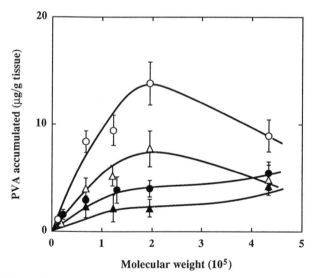

Figure 4. Effect of PVA molecular weight on the accumulation in the carrageenan-induced inflammatory sites (o, ●) and the normal sites (△, ▲) (solid marks: 3 h after injection; open marks: 24 h after injection).

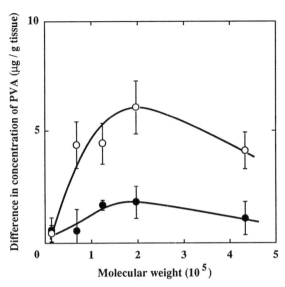

Figure 5. Difference in PVA concentrations between the carrageenan-induced inflammatory and the normal sites, 3 h (●) and 24 h (o) after injection.

Table 1. Effect of the number of carrageenan injections on the PVA accumulation to inflammatory and normal sites (Mw of PVA=196,000)

Injection of carrageenan	Accumulation ratio[a]	
	3 hr	24 hr
Once	1.3 ± 0.2	1.6 ± 0.4
3 times	2.0 ± 0.3	4.2 ± 0.7

a) The ratio of PVA accumulated into inflammatory site to that into normal site.

inflammatory sites to a higher degree compared to the normal sites. Since the effect of PVA molecular weight was not quite clear, the pharmacokinetic analysis was performed for the i.v. injected PVA. Shown in Figure 6 is the accumulation rate at the inflammatory and normal sites, along with the AUC for PVA of different molecular weights after i.v. injection. The PVA accumulation rate decreased with its molecular weight for the both sites. However, the PVA accumulation rate was higher for the inflammatory site than for the normal one, indicating the enhanced vascular permeability caused by the inflammation. Moreover, the difference in the PVA concentrations at the two sites decreased with increased molecular weight. In this regard, the small size PVA appears to be preferable to the larger PVA as the former accumulates in higher concentrations

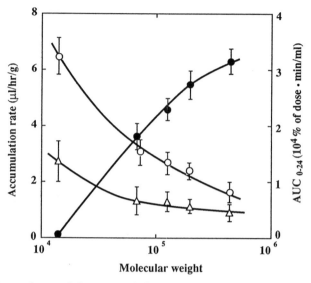

Figure 6. Dependence of the accumulation rate on PVA molecular weight. (o) Rate at the inflammatory site; (Δ) rate at the normal site; (●) AUC_{0-24}.

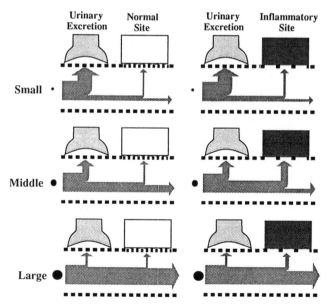

Figure 7. Schematic model for the fate of different Mw PVA's after i.v. injection.

at the inflammatory site. However, the PVA accumulation depends on its concentration in the blood circulation as well. The PVA retention in the blood is proportional to the AUC value which is higher for the PVA with a higher Mw. Thus, the larger sized PVA molecules are retained in the blood for a longer time.

The PVA does not accumulate in the organs, such as liver or spleen, indicating the absence of specific interactions with those cells. On the basis of this information and our experimental data, a model can be proposed regarding the fate of PVA after its i.v. injection. The PVA of different molecular weights are partly urinary excreted, and partly deposited at the inflammatory and normal sites (Figure 7). The main route of the PVA elimination from the blood circulation is by excretion via the kidney glomeruli. The accumulation of the PVA with a lower Mw is higher at the inflammatory site than at the normal one (Figure 6). However, the PVA concentration at the both sites is very low (Figure 5). The small size PVA may be excreted through the kidneys too rapidly for a significant accumulation at the inflammatory site. Furthermore, the localized low Mw PVA is eliminated faster from the tissue than the higher Mw PVA. Indeed, the accumulation rate of the PVA with a higher Mw was low compared with that of the small PVA molecules. On the other hand, the AUC value increased with the increase in the PVA molecular weight. A high AUC value means that the PVA concentration in the tissue is high. Thus, the fate of PVA in the body is affected by the molecular weight as well as the AUC value. A balance of these two factors indicate that the maximum accumulation of the PVA should occur with a medium Mw.

Literature Cited

1. Boddy, A.; Aarons, L. *Advanced Drug Delivery Reviews* **1989**, 3(2), 155-266.
2. Goddard, P. *Advanced Drug Delivery Reviews* **1991**, 6(2), 103-233.
3. Unpublished data.
4. Kamata, R.; Yamamoto, T.; Matsumoto, K.; Maeda, H. *Infect. Immunity* **1985**, 48, 747-753.
5. Illum, L.; Eright, J.; Davis, S.S. *Int. J. Pharm.* **1989**, 52, 221-224.
6. Greenwood, F.C.; Hunter, W.M. *Biochem. J.* **1963**, 89, 114-123.
7. Beauchamp, C.O.; Gonias, S.L.; Menapace, D.P., Pizzo, S.V. *Annal. Biochem.* **1983**, 131, 25-33.
8. Takakura, Y.; Atsumi, R.; Hashida, M.; Sezaki, H. *Pharm. Res.* **1990**, 7, 339-346.
9. Yamaoka, K.; Tanigawara., Y.; Nakagawa, T.; Uno, T. *Pharm. Dyn.*, **1981**, 4, 879-885.

RECEIVED August 23, 1993

Chapter 14

Design of New Building Blocks in Resorbable Polymers

Application in Drug Delivery Microspheres

Ann-Christine Albertsson, Anders Löfgren, Cecilia Sturesson, and Maria Sjöling

Department of Polymer Technology, Royal Institute of Technology, S–100 44 Stockholm, Sweden

Copolymers of 1,5-dioxepan-2-one (DXO) and D,L-dilactide (D,L-LA) were synthesized in different molar compositions. ^{1}H-NMR was used to determine the molar composition of the copolymers, and DSC was used to determine the glass transition and melting temperatures. All the polymers were amorphous with a glass transition varying between -36°C (poly(DXO)) to +55°C (poly(D,L-LA)). *In vitro* hydrolysis studies on the copolymers showed degradation times up to 250 days. Copolymers of 1,3-dioxan-2-one (TMC) and ε-caprolactone (CL) were also studied. Conversion studies were performed, and both monomers separately showed almost the same reactivity. Poly(ε-caprolactone) seemed to be more sensitive to transesterification at elevated temperatures than poly(trimethylene carbonate). Using DSC, melting endotherms were seen in molar compositions with a CL content as low as 65% which indicates a block structure. Microspheres intended for drug delivery were prepared from poly(TMC-co-CL), poly(adipic anhydride) and poly(lactide-co-glycolide). SEM studies showed that the microsphere morphology depended on the polymer concentration at the time of preparation and on the choice of polymer. The drug release profiles showed a dependence on the polymer degradation behavior and on the water penetration into the microspheres.

Tailor-made degradable polymers may be used for different medical applications, such as devices intended for controlled drug release. One advantage of using degradable polymers in drug delivery, suture filaments, ligature clamps, bone fixation plates, and other applications is that the device does not have to be removed after its purpose has been fulfilled *(1)*. The most successful class of degradable polymers so far have been aliphatic polyesters. The degradation takes place via hydrolysis of the ester linkages in the polymer backbone *(2)*. These materials must be extensively tested and characterized since many of them are new

0097–6156/94/0545–0172$08.00/0

structures. This is, however, not sufficient. Equally important is the identification of degradation mechanisms and degradation products. Since high molecular weight polymers are seldom toxic, the toxicity and tissue response after the initial post operative period are related to the compounds formed during degradation. The most important polymer on the market today is poly(lactic acid) (PLA) which upon degradation yields lactic acid, a natural metabolite in the human body *(3)*. The formation of natural metabolites should be advantageous as the body has routes to eliminate them. Other commercial degradable materials are polyparadioxane *(4)*, copolymer of glycolic acid and trimethylene carbonate *(5)* which do not give natural metabolites when degraded. Their most important characteristic is probably the fact that the degradation products are harmless in the concentrations present.

In our laboratory, extensive research is in progress to develop polymers in which the polymer properties can be altered for different applications. The predominant procedure is ring-opening polymerization which provides a way to achieve pure and well defined structures. We have utilized cyclic monomers of lactones, anhydrides, carbonates, ether-lactones, and specifically oxepan-2,7-dione (AA), ß-propiolactone, ε-caprolactone (εCL), 1,5-dioxepan-2-one (DXO), dilactide, and 1,3-dioxan-2-one (TMC) *(6-9)*. The work involves the synthesis of monomers not commercially available, studies of the polymerization to form homopolymers, statistical and block copolymers, development of crosslinked polymers and polymer blends, surface modification in some cases, and characterization of the materials formed. The characterization is carried out with respect to the chemical composition and both chemical and physical structure, the degradation behavior *in vitro* and *in vivo*, and in some cases the ability to release drug components from microspheres prepared from these polymers.

Some parts of our recent work is concerned with the synthesis of copolymers of 1,5-dioxepan-2-one and D,L-dilactide and is presented here. These copolymers were subjected to the *in vitro* degradation. We have also studied the copolymerization of 1,3-dioxan-2-one and ε-caprolactone with the use of coordination type initiators. Along with poly(adipic anhydride) PAA from our laboratory and commercially available poly(lactide-co-glycolide) PLG, this copolymer has been used in the preparation of microspheres for drug delivery. The systems have been studied with respect to morphology and the *in vitro* drug release.

Experimental

Materials

ε-Caprolactone (Merck Schuchart, Germany) was distilled from CaH_2. D,L-Dilactide (Boehringer GmbH, Ingelheim, Germany) was recrystallized twice in dry toluene. Stannous 2-ethylhexanoate (Aldrich Chemical Co., USA), triethyl amine (Merck Schuchart, Germany), BF_3OEt_2 (Aldrich Chemical Co., USA), dibutyl tinoxide Bu_2SnO (Aldrich Chemie, Germany), zinc acetate $ZnAc_2 \cdot 2H_2O$ (Merck Darmstadt, Germany), poly(lactide-co-glycolide) PLG (Boehringer GmbH, Ingelheim, Germany), SPAN 80 and TWEEN 80 (Specialty Chemicals, ICI), sesame oil (Apoteksbolaget, Sweden) and timolol maleate (Sigma Chemical Co., USA) were all used without further purification. Solvents: methylene chloride p.a., chloroform p.a., petroleum ether (b.p. 30-50°C), methanol p.a. and hexane p.a. were used as received.

Monomers. 1,5-Dioxepan-2-one, 1,3-dioxan-2-one, and oxepan-2,7-dione were synthesized as described elsewhere *(6,9,10)*.

Polymers. Poly(adipic anhydride) was prepared by bulk polymerizing oxepan-2,7-dione with triethyl amine as initiator at room temperature for 0.5 h. The polymer was purified by dissolution in methylene chloride and precipitated in cold petroleum ether. The polymer was dried to constant weight *in vacuo* at room temperature.

Polymerization Procedure. The general procedure used in the studies of the polymerization reactions was as follows. A 25 mL serum bottle, containing a teflon-coated magnetic stirrer, was dried at 110°C and used as a reaction vessel. Monomers and initiator were added, and the bottle was sealed with a rubber septum. The bottle was then flushed with inert gas (Ar), immersed in a thermostated oil bath, and kept at constant temperature for the desired reaction time. The polymerization was terminated by cooling the bottle in a refrigerator. The polymer was isolated by dissolution in chloroform or methylene chloride and precipitated in methanol or petroleum ether. The filtration was followed by drying at room temperature *in vacuo*.

In vitro **Hydrolysis.** Hydrolysis studies were carried out using melt-pressed films of the copolymers, 0.5 mm thick. Circular discs with a 13 mm diameter were punched from the films. The discs were immersed in 25 mL serum bottles filled with 20 mL of a pH 7.4 phosphate buffer. The bottles were stored in 37°C without stirring or shaking.

Preparation of Microspheres. The method used was an oil-in-oil solvent evaporation based on the method described by Hyon and Ikada *(11)*. Polymer solutions with concentrations of 5, 10 and 20 (wt/v)% were used and contained 5 wt% of timolol maleate as the drug. The polymer solution was added dropwise to sesame oil, containing 2 wt% SPAN 80, under vigorous stirring. Sonification was used to decrease the particle size. The emulsion was stirred at 50°C for 1 h or at 35°C for 1.5-2 h to evaporate the solvent during the microsphere stabilization. The microspheres were washed 3 to 4 times with hexane. The particles were finally washed with water, containing 0.1 wt% TWEEN 80, to avoid aggregation of the particles and to reduce the initial burst, and then dried *in vacuo*.

In vitro **Drug Release Studies.** The release of timolol maleate from the various microsphere systems was studied under sink conditions by using a pH 7.3 phosphate buffered aqueous medium at 23°C. Batches of the particles were suspended in separate buffer solutions and stirred with magnetic stirrers. Samples were periodically collected, the dispersion was centrifuged and the timolol maleate content in the buffer was determined.

Measurements. To determine the chemical composition, ^{1}H-NMR and ^{13}C-NMR spectra were obtained using a Bruker AC-250 or an AC-400 FT-NMR spectrometer. Samples were dissolved in deuterochloroform (Aldrich Chemical Co., USA) in 5 mm o.d. sample tubes. SEC measurements to determine the molecular weights and distributions were made at 30°C with five μ-styragel columns (500, 10^3, 10^4, 10^5, 100 Å). A Waters model 510 was used with a differential refractometer (Waters 410) as detector. THF or chloroform was used as solvent, with a flow rate of 1.0 mL/min. Polystyrene standards, with a narrow molecular weight distribution of 1.06, were used for calibration. A Copam PC-501 Turbo unit was used for data recording and calculations. For DSC-analysis, a Perkin Elmer DSC-7 with a Perkin Elmer 7700 computer was used. Indium was used as standard for temperature calibration, and the analyses were performed under a constant flow of nitrogen with heating rates of 10 or 20°C/min. The morphology of microspheres was examined by SEM using a Jeol JSM-5400 scanning microscope. The cross-sections of the samples were obtained by cutting the microspheres with a sharp glass edge. The samples were mounted on metal stubs and sputter-coated with gold-palladium (Denton Vacuum Desc II). The drug content and release were assayed spectrophotometrically at 294 nm with a Perkin-Elmer 2, UV/VIS spectrophotometer.

Results and Discussion

Results of Polymerizations. 1,5-Dioxepan-2-one and D,L-dilactide were copolymerized with stannous 2-ethylhexanoate as the initiator at 120°C for 16 h. The molar monomer-to-initiator feed ratio were between 400 and 800 and had no significant effect on the molecular weights which were in the range of 42-70x10^3 g/mol (weight average). Molecular weight distributions were in the range of 1.6-2.0. The molecular weights were determined by SEC relative to polystyrene standards. This is not an absolute method for the Mw determination, but earlier studies have shown that light-scattering measurement data do not differ greatly from the values obtained using the SEC method for these types of copolymers *(9)*. Copolymers with DXO:D,L-LA monomer ratios of 80:20, 50:50, and 20:80 as well as the two homopolymers were synthesized. The compositions of the copolymers were determined by ^{1}H-NMR spectroscopy and were in agreement with the feed ratio for all polymers. Morphologically, the copolymers and homopolymers were totally amorphous based on the absence of melting endotherms in the DSC spectra. The glass transition temperatures obtained were directly dependent on the monomer composition of the copolymers, varying from -36°C (poly(DXO)) to +55°C (Poly(DL-LA)).

Copolymerization of 1,3-dioxan-2-one (TMC) and ε-caprolactone (CL) was studied using stannous 2-ethylhexanoate, dibutyl tin oxide Bu$_2$SnO, and zinc acetate ZnAc$_2\cdot$2H$_2$O as the initiators. All the initiators required a high reaction temperature, typically 80°C or more, to polymerize TMC and CL. The conversion of monomer to polymer was studied for the homopolymerizations. Samples were withdrawn from the reaction vessel at different times and examined with ^{1}H-NMR to determine the molar concentration of the polymer. The temperature was 100°C for the polymerization with Sn-oct, and 120°C for Bu$_2$SnO and ZnAc$_2\cdot$2H$_2$O polymerizations. In the homopolymerizations of TMC with Bu_2SnO and $ZnAc_2 \cdot 2H_2O$, the yields were 100% after 0.3 h, and for Sn-oct after 20 h. Polymerization with Sn-oct is highly dependent on the temperature, and a rise in temperature should increase the polymerization rate. CL showed a high conversion in the Bu_2SnO initiated polymerization. The Sn-oct-initiated polymerization reached 100% conversion after 10 h, and the $ZnAc_2 \cdot 2H_2O$-initiated reaction reached 100% conversion after 20 h.

Organometallic compounds, especially tin compounds, are effective in polymerizing lactones. The slower reaction rate with $ZnAc_2 \cdot 2H_2O$ could be due to the slightly anionic character of the compound. TMC, however, is polymerized by various types of initiators (6). SEC measurements were made to monitor the changes in molecular weight with time. Both polymerizations showed the same behavior. The weight average Mw was in the range of $20-95 \times 10^3$ g/mol, the lowest Mw was obtained with $ZnAc_2 \cdot 2H_2O$, and the highest Mw with Bu_2SnO. The Mw changed with time for Bu_2SnO to 25×10^3 g/mol, which shows the degradation effect of the tin compounds at elevated temperatures. Mw distributions (MwD) were about 2.5, except for the CL polymerizations with Bu_2SnO and $ZnAc_2 \cdot 2H_2O$, where the MwD was 7.6 and 18.5, respectively. The degradation effect is a result of transesterification between the polymer chains. These effects are more pronounced at higher temperature. Copolymers were prepared with 0-100% TMC in the polymer. Similar to the copolymerization of 1,5-dioxepan-2-one and D,L-dilactide, the copolymers had the same ratio between the two monomers as the monomer feed which confirms that the polymerizations were quantitative.

The homopolymer of TMC is amorphous (T_g = -20°C), while poly(caprolactone) is crystalline and melts at 56°C, according to DSC. The melting temperature and the glass transition were recorded for the different copolymers containing 0-100% TMC. The polymers showed a melting endotherm even with a CL content of 65%, and the melting temperature decreased to 29°C. The glass transition could be recorded for polymers containing 38% to 100% TMC, where the former had the transition at -55°C. The copolymers contain fractions consisting of blocks, based on the melting endotherms for the decreased content of CL.

In vitro **Hydrolysis Results.** *In vitro* degradation of copolymers of DXO and LA was followed by characterization of the samples withdrawn from the degradation bottles. The water uptake was evident almost immediately after immersion into the buffer solution. The changes in molecular weight were followed by SEC. In all cases, samples withdrawn after one week showed a lower molecular weight than the original Mw. Shown in Figure 1 is the Mw decrease for the homopolymers, while Figure 2 shows the decrease for the copolymers. When the molecular weight dropped to below 7000 g/mol, it became difficult to make measurements due to small sample mass and solubility difficulties. The highly hydrophilic oligomers tended to clog the precolumn to the SEC-apparatus. The degradation rate varied with the composition of the copolymer; the DXO-rich samples had a slower rate of hydrolysis.

Copolymers of 1,5-dioxepan-2-one and D,L-dilactide are possible candidates for use as a polymer matrix in drug delivery systems. The copolymers are amorphous,

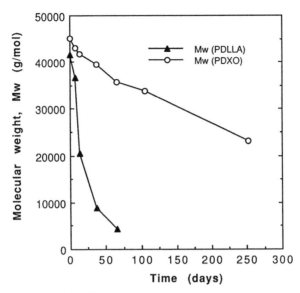

Figure 1. Molecular weight (Mw) as a function of hydrolysis time for DXO and D,L-dilactide homopolymers.

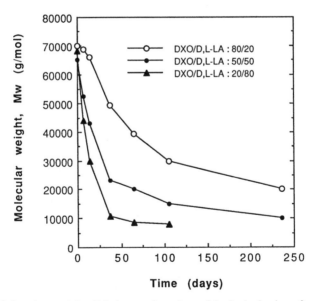

Figure 2. Molecular weight (Mw) as a function of hydrolysis time for copolymers of DXO and D,L-dilactide.

degrade in an uniform way, and the degradation products dissolve completely. The degradation time for oligomers varies over a broad range spanning from years for DXO-rich copolymers to 30-40 days with the D,L-LA rich copolymers. A more extensive study is taking place currently.

Results from Microsphere Preparation. The SEM studies of the microsphere morphology indicate that the roughness and porosity of the surface depended on the polymer concentration in the microspheres and on the choice of polymer. The PLG microspheres with the highest polymer concentration had a very smooth and nonporous surface. As the concentration of polymer was decreased, the surface became uneven. The cross-sections revealed a significant distinction in porosity for the different polymer concentrations used in the preparation of the microspheres. The lowest concentration yielded the most porous interior, as shown in Figure 3.

In contrast to the smooth and relatively dense microspheres obtained with the amorphous PLG, the crystalline PAA microspheres had a very rough surface, porous walls and a hollow center. When these two polymers were blended in a 67:33 (PAA:PLG) molar ratio, the microspheres displayed a surface less rough than that of the PAA microspheres, while for a 20:80 ratio the surface was only dimpled. The porous cross-section for the 67:33 blend revealed a separation of the two polymers which was not observed for the 20:80 blend. The latter microspheres exhibited a large amount of pores with different diameters.

Currently, microspheres prepared from poly(TMC-co-CL) are being examined. These microspheres showed a smooth external surface containing pores. The technique used for cutting the microspheres appeared inadequate for soft matrices like poly(TMC-co-CL) which has a T_g well below room temperature.

In vitro **Release Study Results.** The drug release rate was monitored spectrophoto-metrically. The observed triphasic release behavior and kinetics for the PLG microspheres containing timolol maleate is reported in Figure 4 *(12)*. A SEM study of the microsphere cross-sections revealed significant differences in porosity due to variations in the polymer concentration during preparation. The 5% PLG microspheres exhibited a substantial number of pores facilitating water penetration and drug diffusion. A more rapid hydration of the polymer matrix is assumed to induce an early onset of bulk degradation in the microspheres and a subsequent collapse of the structure. With increasing polymer concentration in the microsphe-res, the cross-sections were found to be increasingly dense. The fact that the porosity is enhanced by a decrease in polymer concentration may explain the differences in release kinetics associated with the changes in PLG concentrations.

In contrast to the PLG system, the PAA system exhibited a very fast drug release: 100% drug was released in 48 h. Taking into consideration the high sensitivity of the aliphatic anhydride to hydrolysis due to its hydrophilic backbone, as well as the porous structure of the microspheres, diffusion-controlled drug release might be expected *(13)*. The drug release appeared to be accompanied with the polymer degradation, and no triphasic release behavior was observed.

The rapidly degrading PAA was blended into the PLG matrix to accelerate the drug release, assuming that pores are formed in the microspheres during the PAA degradation. In the case of a 67:33 (PAA:PLG) blend, the release curve showed no significant deviation from the PAA release curve; 98% drug was released in 48

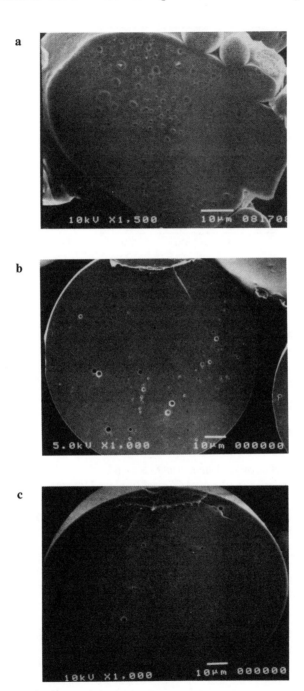

Figure 3. Scanning electron micrographs of cross-sections of PLG microspheres showing the difference in porosity with different polymer concentrations: a) 5% PLG; b) 10% PLG; c) 20% PLG.

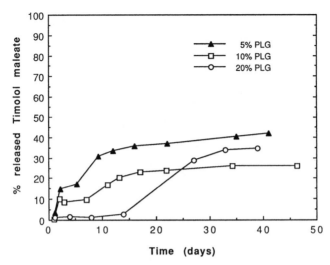

Figure 4. The percentage of timolol maleate released from PLG microspheres as a function of time.

h. When 20% PAA was blended into the PLG matrix, the release rate showed a significant decrease compared to that for the PAA alone, but it was still higher than for the PLG alone. After 24 h, approximately 60% of the timolol maleate was released, and the second phase in the drug release profile for PLG was observed. The release studies for the poly(TMC-co-CL) microspheres are in progress, and the results will be reported later.

Literature Cited

1. Vert, M.; *Angew. Makromol. Chem.* **1989**, 166-167, 155.
2. Gilding, D.K., In *Biocompatibility of Clinical Implant Materials*, vol. II; Williams, D.F. Ed.; CRC Press Inc., Boca Raton, Florida, 1981, p 218.
3. Kulkarni, R.K.; Pani, K.C.; Neuman, C.; Leonard, F. *Arch. Surg.* **1966**, 93, 839.
4. Doddi. N.; Versfeldt, C.C.; Wasserman, D., US patent 4,052,988, 1977.
5. Casey, D.J.; Roby, M.S., Eur. Patent EP 098 394 A1,1984.
6. Albertsson, A.-C.; Sjöling, M. *J. Macromol. Sci., Chem.* **1992**, A29, 43.
7. Mathisen, T.; Masus, K.; Albertsson, A.-C. *Macromolecules* **1989**, 22, 3842.
8. Lundmark, S.; Sjöling, M.; Albertsson, A.-C. *J. Macromol. Sci., Chem.* **1991**, A28, 15.
9. Albertsson, A.-C.; Löfgren, A. *Makromol. Chem., Macromol. Symp.* **1992**, 53, 221.
10. Albertsson, A.-C.; Lundmark, S. *J. Macromol. Sci., Chem.* **1990**, A27, 397.
11. Hyon, S.-H.; Ikada, Y., Eur. Patent EP 330 180, 1989.
12. Sturesson, C.; Carlfors, J.; Edsman, K.; Andersson, M. *Int. J. Pharm.* **1993**, 89, 235-244.
13. Mathiowitz, E.; Amato, C.; Dor, Ph.; Langer, R. *Polymer* **1990**, 31, 547.

RECEIVED August 9, 1993

Chapter 15

Chain Structure and Hydrolysis of Nonalternating Polyester Amides

Kenneth E. Gonsalves and Xiaomao Chen[1]

Polymer Science Program, Institute of Materials Science, U–136, and Department of Chemistry, University of Connecticut, Storrs, CT 06269

Two series of polyesteramides, PEA-I and PEA-II, were synthesized and subjected to hydrolytic degradation. PEA-I was made by a two-step polycondensation reaction from 1,6-hexanediol, adipoyl chloride and 1,6-hexanediamine, while PEA-II was synthesized via the anionic copolymerization of ε-caprolactam and ε-caprolactone. Based on selective ester hydrolysis by base treatment, both types of polyesteramides proved to have random chain structures. The hydrolysis of these polyesteramides in buffer solutions was greatly enhanced with the increase of temperature and ester content of the copolymers. Both acid and base were found to catalyze the hydrolysis reaction of the copolymers.

Various aliphatic polyesteramides containing α-hydroxy or α-amino acid moieties, have been studied for their potential biomedical applications *(1-7)*. Polyesterpeptides, which are a class of alternating polyesteramides of α-hydroxy acid and α-amino acid, were reported to be hydrolyzable and biodegradable *(1-4)*. Helder *et al. (5)* studied the kinetics of *in vitro* degradation of nonalternating glycine/DL-lactic acid copolymers. Barrows and coworkers *(6,7)* synthesized a series of alternating polyesteramides by using a two-step polycondensation reaction from glycolic acid, diamine and diacyl chloride, and investigated the *in vivo* degradation of the copolymers for the purpose of developing new surgical implants. However, very few reports on the degradation of polyesteramides, which might be potentially useful for more general purposes such as packaging, were found in the literature. Tokiwa *et al. (8,9)* showed that polyesteramides, made by an amide-ester interchange reaction of nylons and polycaprolactone at 270 °C, were degraded by the enzyme lipase. To develop degradable materials for potential biomedical and other applications, two types of nonalternating PEAs, containing no α-hydroxy or α-amino acid moieties, were synthesized by interfacial and anionic polymerizations.

[1]Current address: Department of Chemical and Biochemical Engineering, University of Iowa, Iowa City, IA 52242

0097–6156/94/0545–0181$08.00/0

The initial results of the characterization, and hydrolytic degradation of these nonalternating copolymers is presented.

Experimental

Synthesis. PEA-I copolymers were made by a two-step condensation polymerization from 1,6-hexanediamine, adipoyl chloride and 1,6-hexanediol. The method of synthesis has been reported previously *(10)*. PEA-II copolymers were made by the anionic ring-opening copolymerization of ε-caprolactam and ε-caprolactone at 100° to 160°C with N-sodiocaprolactam as a catalyst *(11a)*.

Selective Hydrolysis of Ester Groups. PEA powder (2.0 g) was placed in a flask containing a solution of potassium hydroxide (2.5 g) in 10 mL water and 40 mL ethanol. The suspension was kept at 35°C for 6-8 h with stirring, until all the ester bonds were hydrolyzed as indicated by FTIR spectra (when the absorption of ester $C=O$ at 1730 cm^{-1} had disappeared). After the hydrolysis, ethanol was removed at reduced pressure and 20 mL distilled water was added to the residue. The mixture was adjusted to pH 3 with 1 N HCl. The insoluble solid (**part-A**) was separated by filtration, washed thoroughly with distilled water and then dried in a vacuum oven for 2 days at 40°C. The filtrate (combined solution) was dried in vacuum and extracted with methanol. An oily product (**part-B**) was obtained after the solvent was removed in vacuum.

Carboxylic acid end group analysis of **part-A** was conducted as follows: Under the flow of argon, 0.2 g solid sample was dissolved in 8 mL of benzyl alcohol at elevated temperature (165°C for samples hydrolyzed from PEA-I, 110°C for the PEA-II samples). The solution obtained was titrated with warm 0.05 N NaOH solution of benzyl alcohol to phenolphthalein (in 1:1 methanol/water) end point, while it was hot. The blank without polymer was also titrated under the same conditions. The 0.05 N NaOH solution was calibrated with standard 0.05 N benzoic acid solution in benzyl alcohol, where phenolphthalein was used as an indicator.

Techniques

Intrinsic Viscosities. The intrinsic viscosities of polymer samples were measured from a 0.5 g/dL solution of 90% formic acid at 25°C.

FTIR Spectroscopy The infrared spectra of the polymer films were recorded with a Nicolet 60SX FTIR spectrometer. The films were made by casting solutions of the polymers on NaCl (from methanol solution) or CaF$_2$ (from formic acid or methanol/water solutions) crystal plates, followed by drying in vacuum.

^{1}H-NMR Spectroscopy. ^{1}H-NMR spectra of polymers were obtained on an IBM AF-270 NMR Spectrometer (270 MHz), where CF$_3$COOD was used as a solvent for polyesteramide samples and CDCl$_3$ for polyester oligomer samples.

Thermal analysis. Thermal analysis of the polymers was performed using a Perkin-Elmer DSC-7 (differential scanning chromatograph) in N_2 at a heating rate of 20°C/min.

GPC Analysis. Molecular weight (Mw) and Mw distributions were measured using a Waters 150-C ALC/GPC equipped with m-Styragel HT columns of 10^4, 10^3 and 10^3 pore sizes at flow rate of 1 mL/min. m-Cresol and DMAC (dimethylacetamide) were used as solvents at 100° and 70°C, respectively. Polystyrene Mw standards were used for calibration. The universal calibration method was applied.

Tensile Property Measurement. The polymer films were prepared by melt-molding at 10-20°C beyond their melting points for 3 min, followed by cooling with tap water. Fibers of RA40 were made by melt-spinning at 260°C with a diameter about 0.1 mm. Tensile tests were made using an Instron Model 1101 at a speed of 1 in/min.

Hydrolysis. The hydrolysis of the copolymers was performed according to the following procedure: a piece of polymer film (0.7 mm thick, 100 mg) was placed in a vial containing 10 mL of a buffer solution (pH 10.5, 7.4 or 4.4) with 0.03% sodium azide to inhibit bacterial growth *(5)*. The vial was maintained at a certain temperature (20, 37, or 55°C) and the buffer solution was refreshed once the change of pH value was larger than 1.0.

Samples were removed from the solutions at various times, washed with distilled water, and dried in vacuum to constant weights. The Mw's of these samples were measured by GPC.

Results and Discussion

Study of Chain Structure by Selective Hydrolysis

General Properties. The compositions, [η] (intrinsic viscosity), and DSC data of PEA-I and PEA-II copolymers are shown in Table I. The moderately high values of intrinsic viscosities of PEA-I copolymers indicate that they have relatively high molecular weights. All PEA-I copolymers showed a polyamide melt transition, while a polyester transition appears in the ester-rich copolymer (RA50). FTIR showed very similar IR spectra for all four PEAs except that the relative intensities of the absorptions of the COO and CONH groups varied with the chemical composition. Absorptions were observed at 3306 and 3056 cm^{-1} (N-H stretching of amide), 1737 and 1637 cm^{-1} (C=O stretching of the ester group and amide-I bands, respectively), and 1534 cm^{-1} (amide-II band). Similar results were observed by Castaldo *et al. (12)* for analogous polyesteramides.

PEA-II copolymers possess a random microstructure, with chains composed of -NH(CH$_2$)$_5$CO- (lactam) and -O(CH$_2$)$_5$CO- (lactone) groups *(10)*. Each of the copolymers displayed a single melting point on DSC analysis. The IR spectra showed typical polyesteramide absorptions, similar to those of PEA-I copolymers.

Table I. Composition, Intrinsic Viscosities, and DSC Data of PEA Copolymers

Sample	PEA-I			PEA-II			
	RA25[a]	RA40	RA50	AE45	AE55	AE70	AE90
$[\eta]$, dL/g	1.09	1.30	0.57	0.26	0.35	0.34	0.74
Amide content (mol%):							
feed ratio	75	60	50	45	55	70	90
found by ^{1}HNMR	58	54	34	31.8	44.3	56.7	87.7
DSC Data:							
T_m, °C	255.7	252.2	247.0[c]	46.7	89.0	115.7	205.4
ΔH, J/g	41.34	15.64	12.88	27.3	24.9	-	43.6

[a]RA25 represents random **PEA-I** copolymer made from adipoyl chloride with 2 5 % of ester, while AE55 represents **PEA-II** copolymer with 55 % of amide.
[b]Amide % were calculated from the peak integration ratios of methylene groups next to -COO- and -CONH- groups.
[c]Another peak was observed at 50.0°C with ΔH of 19.2 J/g.

A detailed characterization of this type of polyester-amides has been reported previously (10,11).

It is interesting to note that PEA-I copolymers possess a polyamide transition or both polyamide and polyester transitions, while the melt transitions of PEA-II and alternating polyesteramides (10,11a) fall between those of the corresponding polyamide and polyester. It was unexpected that a random copolymer like PEA-I would show the typical homopolymer transition, instead of an intermediate transition. To explain this, the following hypothesis is proposed: the amide-ester interaction in PEA-I copolymers is weak compared to amide-amide and ester-ester interactions, so that the amide and ester segments tend to dislike each other and form micro-crystalline domains separately. The T_g for PEA-II copolymers have been reported by Goodman et. al. (11). In general, the T_g of PEA-II type copolymers increased with increasing amide content, and fell in the range -50°C and +50°C. The T_g of AE-70 and AE-90 were -27.3°C and +13.1°C respectively. It was also observed by us that the Tg decreased by 20-30°C after the polymer had been saturated with water. A T_g for the PEA-I copolymers could not be obtained by DSC, possibly due to the interference of the melting peak of the ester block in the same region. (T_g's of -10°C to 30°C have been reported for analogous polymers (12).

Work-up of Hydrolysis. To elucidate the chain structure of the polyesteramides, hydrolysis of the ester groups of the copolymers with base was conducted. After the powder samples of copolymers were treated with a solution of KOH in ethanol/H_2O at 35°C for 6-8 h, only the ester groups were hydrolyzed and the amide groups were unaffected. The IR spectra of all samples showed that the hydrolysis of ester bonds (1730 cm^{-1}) was complete after base treatment for 6 h under the given conditions.

The base hydrolysis of PEA-I and PEA-II copolymers is outlined in Schemes 1 and 2, respectively. For PEA-I, the final products (after acidifying to pH 2-3) would be hexanediol-1,6, adipic acid and oligoamides of various chain length with carboxylic acid as end groups on both sides (Scheme 1). The solid portion (**part-A**) consisted of the oligoamide mixture, whereas the liquid portion (**part-B**) consisted of 1,6-hexanediol and adipic acid (**part-B** might contain some small water-soluble oligoamides). For PEA-II, **part-A** would contain various oligoamides with -OH and -COOH as end groups on either ends, whereas **part-B** (water soluble) would contain 6-hydroxycaproic acid and some short chain oligoamides (DP = 1 or 2).

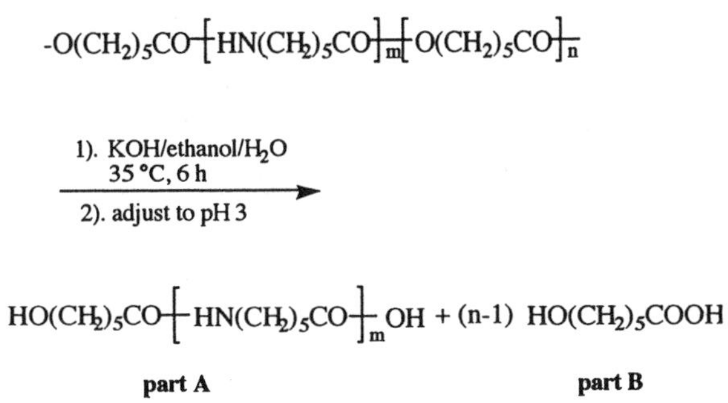

$$-O(CH_2)_5CO\!\!\left[HN(CH_2)_5CO\right]_m\!\!\left[O(CH_2)_5CO\right]_n$$

1). KOH/ethanol/H_2O
   35 °C, 6 h
———————————————→
2). adjust to pH 3

$$HO(CH_2)_5CO\!\!\left[HN(CH_2)_5CO\right]_m\!\!OH + (n\text{-}1)\ HO(CH_2)_5COOH$$

part A   **part B**

Scheme 1. Selective hydrolysis of PEA-I copolymers.

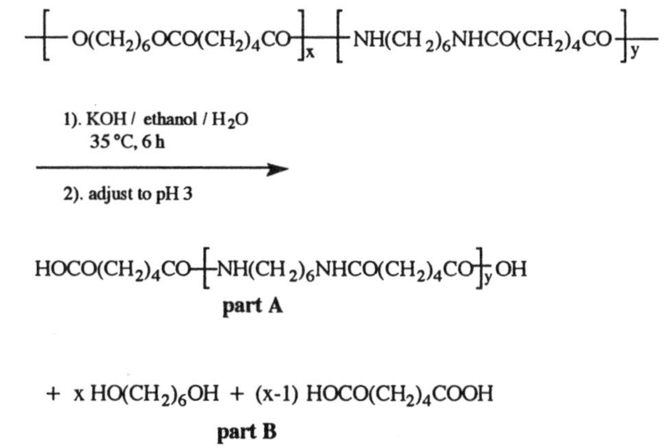

$$\left[O(CH_2)_6OCO(CH_2)_4CO\right]_x\!\!\left[NH(CH_2)_6NHCO(CH_2)_4CO\right]_y$$

1). KOH / ethanol / H_2O
   35 °C, 6 h
———————————————→
2). adjust to pH 3

$$HOCO(CH_2)_4CO\!\!\left[NH(CH_2)_6NHCO(CH_2)_4CO\right]_y\!\!OH$$
part A

$$+\ x\ HO(CH_2)_6OH + (x\text{-}1)\ HOCO(CH_2)_4COOH$$
part B

Scheme 2. Selective hydrolysis of PEA-II copolymers.

Analysis of Soluble Portion (Part-B). The IR spectra of **part-B** of PEA-I copolymers showed no absorptions of amide-I and II at 1520-1650 cm^{-1}. ^{1}H-NMR spectra also showed no amide peaks in these solids, as indicated by the absence of peaks at δ 2.3-2.7 (C-CH$_2$-CO-N) and 3.2-3.5 (N-CH$_2$-C). Thus, it was concluded that all the oligoamides hydrolyzed from PEA-I copolymers were insoluble in water. This could also explain the fact that more product was obtained in **part-A** with the higher amide content in the copolymer, as shown in Table II. For PEA-II copolymers, the soluble portion contained some small oligoamides (DP $\leq$ 3) which are soluble in water *(11c)* together with 6-hydroxycaproic acid, as proven by FTIR and ^{1}H-NMR measurements.

Analysis of Insoluble Portion (Part-A). As shown in Table II, the amount of **part-A** increased with increasing amide content of original copolymer. These solids showed typical oligoamide characteristics. Only traces of ester groups and hydroxyl end groups remained in **part-A**, as indicated by IR and ^{1}H-NMR spectra of the hydrolysate. Thus, the molecular weights of the oligoamides could be estimated from end-group analysis.

Two methods (titration and ^{1}H-NMR) were used for end group analysis. The M_n values obtained from these analyses are listed in Table III, based on the assumption that both end groups were -COOH for PEA-I, and the end group was -COOH on one side and -OH on the other side for PEA-II.

As mentioned above, the hydrolysis of the ester groups was nearly complete and no appreciable breakdown of amide bonds was observed under the given conditions. Therefore, the oligoamides recovered from **part-A** would be the amide segments of the original copolymers. Together with the M_n values of the oligoesters, it is possible to determine the chain structure of these types of copolymers.

Table II. Selective Hydrolysis of Polyesteramides

Sample	AE45	AE70	AE90	RA40
Weight (original), g				
	4.06	2.03	2.04	2.06
Weight (after)[a], g:				
Part-I (insoluble)	0	0.70	1.98	1.29
Part-II (soluble)	4.28	1.28	0.08	0.85

[a]The weight was not calibrated for the end groups (-OH, -COOH) of the products (this may be the reason for the weight imbalance).

Table III. Analysis of the Solid Portion after Hydrolysis of PEA

Sample	AE70	AE90	RA40
T_m (DSC), °C	118.6	199.8	157.6
ΔH, J/g	76.8	83.2	49.9
[OH⁻][b], mmol/g	1.64	0.514	0.218
[-COOH], mmol/g			
titration	1.68	0.634	2.34
¹H-NMR	1.94	0.714	2.93
M_n, g/mol			
titration	595	1577	885
¹H-NMR	559	1628	682

[a]Another weak peak was observed at 229.0°C with ΔH of 17.9 J/g.
[b]By ¹H-NMR.

Hydrolysis Studies of PEA-I Copolymers. The water absorption of PEA-I copolymers was 3-5%, measured after immersing the film (0.7 mm thick) in water overnight. It was between that for polyesters (< 3%) and polyamides (5-8%). The hydrolysis of RA25 and RA40 (0.7 mm thick films) was conducted in pH 7.4 buffer solutions at 37°C. The changes of molecular weights and the weight losses were gradual (Table IV) with time. The slow ongoing hydrolytic degradation of this type of polyesteramide is evident from the data listed in Table IV.

RA40 was melt-spun into fibers at 260°C with a diameter of about 100 mm. These non-oriented fiber samples were then subjected to hydrolysis in pH 7.4 buffer solution at 22 and 37°C, respectively. After treatment, the tensile properties of the fiber were measured to assess the degradation of the sample. The results of elongation and tensile strength, as a function of the time of hydrolysis at 22 and 37°C, are shown in Figure 1. The tensile strength and the elongation of RA40 fiber remained the same after hydrolysis at 22°C for 80 days, while both properties of this fiber decreased dramatically after hydrolysis at 37 °C for the same period of time.

Hydrolysis of PEA-II Copolymers

Water Absorption. The water absorption of PEA-II copolymers, measured by

Table IV. Weight Loss and GPC Data of PEA-I Copolymers after Hydrolysis
(pH 7.4, 37°C)

Time (days)	RA25		RA40	
	% wt. loss	M_w/M_n	% wt. loss	M_w/M_n
0	0	1.695	0	7.516
36	3.1	1.741	3.6	8.038
160	6.0	data not available	7.2	6.544
242	6.3	1.741	11.5	data not available

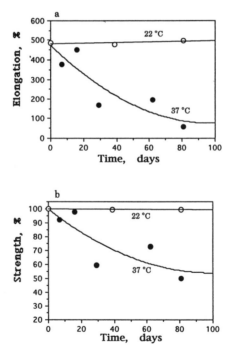

Figure 1. Tensile properties of RA40 fiber vs. hydrolysis in pH 7.4 buffer.
(a) Elongation, (b) Tensile strength.

immersing the polymer films in distilled water at 22°C, are within 1-7%. It was also observed that, for all copolymers, the water content increased with increasing amide content.

PEA-II copolymers (amide content ≥ 55%) possess good mechanical properties *(10,11)*. However, except for AE90, the mechanical properties such as tensile elongation of these copolymers will be lost completely (i.e., elongation was nearly 0 %) after they have been immersed in water for longer than one day, before any chemical degradation occurs. The surprisingly poor properties of these copolymers in the wet state might be related to: a) the relatively low molecular weight of the copolymers (low $[\eta]$ in Table I) compared to most commercial nylons; b) lower crystallinity; and c) the elimination of the intermolecular hydrogen bonds of the polymers by the absorbed water in the bulk.

Degradation Profile. The hydrolysis of PEA-II copolymers in buffer solutions has been studied in detail to observe the effects of pH, temperature and composition on the degradation of polyesteramides. The polymer films (0.7 mm thick) were immersed in buffer solutions with various pH values (pH 10.5, 7.4, and 4.4), at different temperatures (20, 37 and 55°C). The weight and GPC measurements showed that the copolymers degraded under these conditions, as seen from Figures 2 to 5.

It is apparent that the degradation of AE55 and AE70 was greatly enhanced by increasing the temperature from 37°C to 55°C. As shown in Figures 3a and 5a, the weight loss was faster in a pH 10.5 (basic) buffer solution than that in a pH 4.4 (acidic), and 7.4 (neutral) buffer solution for both AE55 and AE70. On the other hand, the reduction of molecular weight was greatly enhanced in acidic medium (pH 4.4) compared to that in neutral medium (pH 7.4), as observed from the data in Figures 2b and 4b. These results indicate that the hydrolysis of the copolymers occurred at increased rates in both basic and acidic media compared to that in neutral medium.

Comparing the degradation profiles of AE55 with those of AE70, it was observed that AE55 degraded at a much faster rate than AE70. This indicates that the hydrolytic degradation of polyesteramides is enhanced by increasing the ester content of the copolymer.

The polymer composition with degradation time was monitored by ^{1}HNMR spectroscopic analysis of the remaining part of the polymer film. The relative ester content of the copolymer decreased on hydrolysis, while the amide content and the hydroxyl end group increased, as shown in Table V.

Mechanism of Hydrolysis. The copolymers of ε-caprolactam and ε-caprolactone may be assumed to degrade via amide bond scission, ester amide scission or both. As found in this work by weight loss and GPC analyses, no appreciable degradation was observed for nylon 6 (a typical polyamide), except when it was subjected to hydrolysis in acidic solution at high temperature for a relatively long time (pH 4.4, 55°C, 140 days). It has also been shown that the *in vivo* degradation of alternating polyesteramides results from the hydrolysis of ester bonds *(6,7)*. Based on these

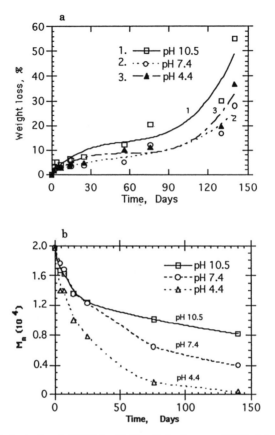

Figure 2. Hydrolysis of AE55 in Buffer Solutions at 55°C. (a) % wt. loss; (b) Mw change.

considerations, it is suggested that the amide bond scission is less important for the hydrolytic degradation of the copolymers under the given conditions. Thus, the hydrolytic degradation of these copolymers is mainly via the ester bonds.

The hydrolytic degradation of polyesters is generally attributed to the hydrolysis of the ester bond to produce carboxylic acid and alcohol, which is catalyzed by either acid or base:

$$\text{---CH}_2\text{COOCH}_2\text{---} \quad + \text{ H}_2\text{O} \xrightarrow{\quad \text{H}^+ \text{ (or OH}^-\text{)} \quad} \text{---CH}_2\text{COOH} + \text{---CH}_2\text{OH}$$

There is no doubt about the catalytic role of H^+ (proton) in the hydrolysis of polyesters. However, the insensitivity of polyester degradation rate to the pH of the reaction medium has been reported *(13-15)*. To explain this, a mechanism

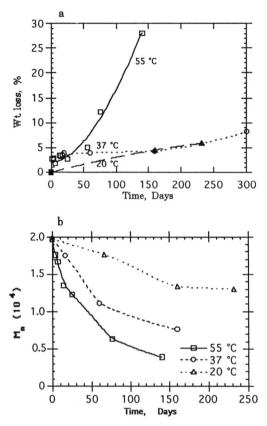

Figure 3. Hydrolysis Profile of AE55 in pH 7.4 buffer. (a) % wt. loss; (b) Mw change

was proposed which is related to an acidic microenvironment resulting from carboxylic acid groups generated by the ester bond cleavage *(16)*. Nevertheless, the effect of pH on the hydrolysis of polymers containing ester groups needs to be further clarified. In this work, it was also found that the rate of the degradation of polyesteramides is enhanced in acidic medium. However, as observed from Figures 2a and 4a, the weight loss of the copolymers is favored in basic medium (pH 10.5). This can be explained by the hypothesis that a carboxylic salt will be formed by the ester chain cleavage in basic medium, and that those products will be more soluble in water than the corresponding carboxylic acid products in neutral and acidic media. Combining the effects of both weight loss and decrease of molecular weight, the hydrolysis of the copolymers might be catalyzed by both H^+ and OH^-.

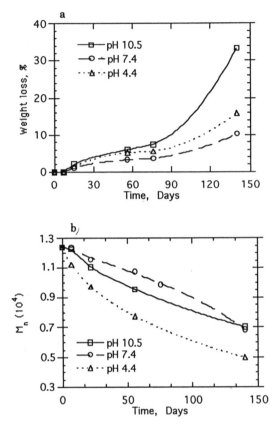

Figure 4. Hydrolysis of AE 70 in Buffer Solutions at 55°C. (a) % wt. loss, (b) Mw change.

Conclusions

Both types of nonalternating polyesteramides were found to be degradable in buffer solutions. For PEA-I copolymers, the degradation has been probed by weight loss, the decrease of molecular weight and decay in mechanical properties. The degradation of PEA-I copolymers was found to be very slow at room temperature and greatly enhanced at 37°C.

It was found that higher temperatures and higher ester content in the PEA-II copolymer produced faster hydrolytic degradation. The weight loss is favored in basic medium compared to acidic or neutral media. It was also observed that the degradation was more rapid in both acidic and basic media than in neutral medium.

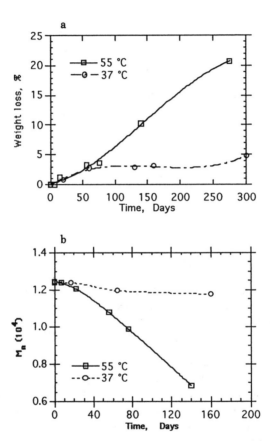

Figure 5. Hydrolysis Profile of AE70 in pH 7.4 Buffer. (a) wt. loss %, (b) M_n changes.

Table V. Relative Composition (by ^{1}H-NMR)^a of AE70 (PEA-II) vs. Hydrolysis (55°C, 150 days)

	Original	pH 4.4	pH 7.4	pH 10.5
Ester	0.4066	0.3585	0.3815	0.3498
Amide	0.5934	0.6415	0.6185	0.6501
Hydroxyl	0.0141	0.0471	0.0344	0.0373
% wt. loss	0	36.6	20.7	49.4
Mw	24,327	7,868	11,561	11,790
M_n	12,408	5,010	6,484	7,020
Mw/M_n	1.961	1.571	1.688	1.679

[a] ^{1}H-NMR spectroscopy was applied to semiquantitatively assess the relative concentration of amide, ester and hydroxyl groups by determining the peak integration of the methylene groups next to them *(17)*.

Literature Cited

1. Asano, M.; Yoshida, M.; Kaetsu, Katakai, R.; Imai, K.; Mashimo, T.; Yuasa, H.; Yamanaka, H. *J. Jpn. Soc. Biomat.* **1985**, 3, 85.
2. Asano, M.; Yoshida, M.; Kaetsu, Katakai, R.; Imai, K.; Mashimo, T.; Yuasa, H.; Yamanaka, H. *Seitai Zairyo* **1986**, 4, 65.
3. Kaetsu, I.; Yoshida, M.; Asano, M.; Yamanaka, H.; Imai, K.; Yuasa, H.; Mashimo, T.; Suzuki, K.; Katakai, R.; Oya, M. *J. Controlled Rel.* **1986**, 6, 249.
4. Schakenraad, J.M.; Nieuwenhuis, P.; Molenaar, I.; Helder, J.; Dijkstra, P.J.; Feijen, J. *J. Biomed. Mater. Res.* **1989**, *23*, 1271.
5. Helder, J.; Dijkstra, P.J.; Feijen, J. *J. Biomed. Mater. Res.* **1990**, 24, 1005.
6. Barrows, T.H.; Johnson, J.D.; Gibson, S.J.; Grussing, D.M. In *Polymers in Medicine II*, Chiellini, E., Giusti, P.,Migliaresi, C., Nicolais, E., Eds.; Plenum Press: New York, 1986; p.85.
7. Horton, V.L.; Blegen, P.E.; Barrows, T.H.; Quarfoth, G.L.; Gibson, S.J.; Johnson, J.D.; McQuinn, R.L. In *Progress in Biomedical Polymers*, Gebelein, C.G., Dunn, R.L., Eds.; Plenum Press: New York, 1990; p.263.
8. Tokiwa, Y.; Suzuki, T.; Ando, T. *J. Appl. Polym. Sci.* **1979**, *24*, 1701.
9. Tokiwa, Y.; Ando, T.; Suzuki, T.; Takeda, K. In *Agricultural and Synthetic Polymers, Biodegradation and Utilization, (ACS Series 433)* Glass, J. E.; Swift, G., Eds.; Am. Chem. Soc.: Washington D. C., 1990; Chapter 12, p. 137.
10. Gonsalves, K.E.; Chen, X.; Cameron, J.A. *Macromolecules* **1992**, *25*, 3309.

11. a) Goodman, I.; Vachon, R.N. *Eur. Polym. J.* **1984**, 20, 529; b) *ibid.*, **1984**, *20*, 539; c) *ibid.*, **1984**, 20, 549.
12. Castaldo, L.; Candia, F.D.; Maglio, G.; Palumbo, R.; Strazza, G. *J. Appl. Polym Sci.* **1982**, 27, 1809.
13. Reed, A.M.; Gilding, D.K. *Polymer* **1981**; 22, 494.
14. Pitt, C.G.; Chaslaw, F.I.; Hibionada, Y.M.; Klimas, D.M.; Schindler, A. *J. Appl. Polym. Sci.* **1981**, 28, 3779.
15. Chu, C.C. *J. Biomed. Mater. Res.* **1985**, 15, 19.
16. Kenley, R.A.; Lee, M.O.; Mahoney, T.R.; Sanders, L.M. *Macromolecules* **1987**, 20, 2398.
17. Gonsalves, K.E.; Chen, X., *Polym. Prepr.* **1992**, 33, 53.

RECEIVED July 29, 1993

Chapter 16

Polymeric Site-Directed Delivery of Misoprostol to the Stomach

P. W. Collins[1], S. J. Tremont[2], W. E. Perkins[3], R. L. Fenton[2], M. P. McGrath[2], G. M. Wagner[2], A. F. Gasiecki[1], R. G. Bianchi[3], J. J. Casler[3], C. M. Ponte[3], J. C. Stolzenbach[4], P. H. Jones[1], and D. Forster[2]

[1]Department of Chemistry, G.D. Searle & Company, Skokie, IL 60077
[2]Monsanto Corporate Research, St. Louis, MO 63198
Departments of [3]Immunoinflammatory Diseases and [4]Drug Metabolism, G.D. Searle & Company, Skokie, IL 60077

Misoprostol is a synthetic 16-hydroxy analog of natural prostaglandin E_1 and is used for prevention of gastric ulcers caused by non-steroidal anti-inflammatory drugs. Misoprostol exerts its therapeutic effect when applied directly to the gastric mucosa. The undesired side effects of misoprostol are observed for the blood borne drug (uterotonic), or, in the case of diarrhea, combined systemic and intestinal exposure. A strategy to reduce side effects while maintaining therapeutic efficacy of the drug is controlled slow delivery of misoprostol to the stomach. A number of functionalized polymer-based systems designed to slowly release misoprostol in the stomach but not in the intestines have been developed. The key polymer-bound release mechanism is a covalent silicon ether bond to the C-11 hydroxy group of misoprostol. The silyl linker releases intact misoprostol from the polymer matrix under acidic conditions (pH 1-3) but not at the higher intestinal pH (>5).

Misoprostol is a synthetic 16-hydroxy analog of natural prostaglandin E_1 (PGE_1). Misoprostol possesses both gastric antisecretory and mucosal protective properties. At present, misoprostol is approved and being marketed in the U.S. for prevention of gastric ulcers caused by non-steroidal anti-inflammatory drugs. In general, misoprostol is a safe and well tolerated drug. However, it is contraindicated in pregnant women because of its uterotonic activity. It also causes mild diarrhea and abdominal discomfort in some patients (10% in clinical studies) *(1)*. In addition, four times a day dosing is recommended due to a short duration of misoprostol activity.

Misoprostol and other gastric antisecretory/mucosal protective prostaglandins exert their therapeutic effect when applied directly to the gastric mucosa *(2)*. The side effects are caused by the presence of the drug in the blood circulation (uterotonic activity), or a combination of intestinal and systemic exposure

0097–6156/94/0545–0196$08.00/0

Misoprostol

(diarrhea). Thus a logical strategy for maintaining antiulcer efficacy while reducing side effects is to effect controlled, slow delivery of the drug to the stomach, to provide sufficient local therapeutic levels, and to limit systemic and intestinal exposure to the drug. The prolonged gastric availability of the drug from the delivery system might also reduce the dosing frequency.

With these criteria in mind, a number of polymer-based delivery systems for misoprostol have been developed and studied *(3)*. These systems were designed to provide a slow release of misoprostol in the stomach but not in the intestines. The key element of the polymer-bound mechanism for selective gastric release is a covalent silicon ether bond to the C-11 hydroxy group of misoprostol. The silyl linker releases intact misoprostol from the polymer matrix under acidic conditions (pH 1-3) but not at the higher intestinal pH values (>5). The steric and electronic nature of the substituents on the silicon linkage has a profound effect on the release rate of the drug in acidic media. Thus, in the present work we prepared and evaluated a series of systems containing different silyl substitution patterns (Table I),

Experimental Methods

Synthesis

The polymeric delivery systems in Table I were prepared in six steps from commercially available polybutadiene (Aldrich) (Figure 1) *(3)*. The six steps were: 1) hydroformylation of polybutadiene; 2) reductive amination of the polyaldehyde to the polyamine; 3) hydrosilylation of the polyamine; 4) coupling misoprostol to the silyl chloride polymer; 5) crosslinking of the miso-polymer; 6) methylation of the crosslinked polyamine.

The first step, hydroformylation of polybutadiene, is the key step in controlling the functional density on the polymer. It was executed at 80°C and 2172 kPa with a 1:1 carbon monoxide/hydrogen mixture in the presence of a catalytic amount of hydridocarbonyl tristriphenylphosphine rhodium and excess triphenylphosphine *(4)*. The reaction progress was monitored by gas uptake from a calibrated small reservoir. The reaction was terminated when 33.5% of the double bonds of the polybutadiene were hydroformylated, which was confirmed by ^{1}H-NMR.

The polyaldehyde was converted to the dimethylamino functionalized polymer by reductive amination with dimethylamine. The polymer product was

Table I. Comparative Pharmacology of Misoprostol Polymer Systems

Polymer-Si(R_1R_2)-O-Misoprostol

System	Silicon Substitution	ED_{50} ($\mu g/kg$ i.g.)	
		Antiulcer Activity (Indo/Gastric)	Diarrheagenic Activity
1	$R_1 = R_2 = Me$	9.7	715*
	Miso/HPMC	13.2	265
2	$R_1 = R_2 = Et$	7.7	910*
	Miso/HPMC	9.5	290
3	R_1 = Phenyl, $R_2 = Et$	7.0	858*
	Miso/HPMC	7.6	325
4	$R_1 = R_2 =$ Phenyl	16.0	2031*
	Miso/HPMC	27.1	485
5	$R_1 = R_2 =$ Isopropyl	32.6	No diarrhea at 1865;
	Miso/HPMC	20.5	430

*Significantly greater than control (misoprostol/HPMC), $p < 0.05$.

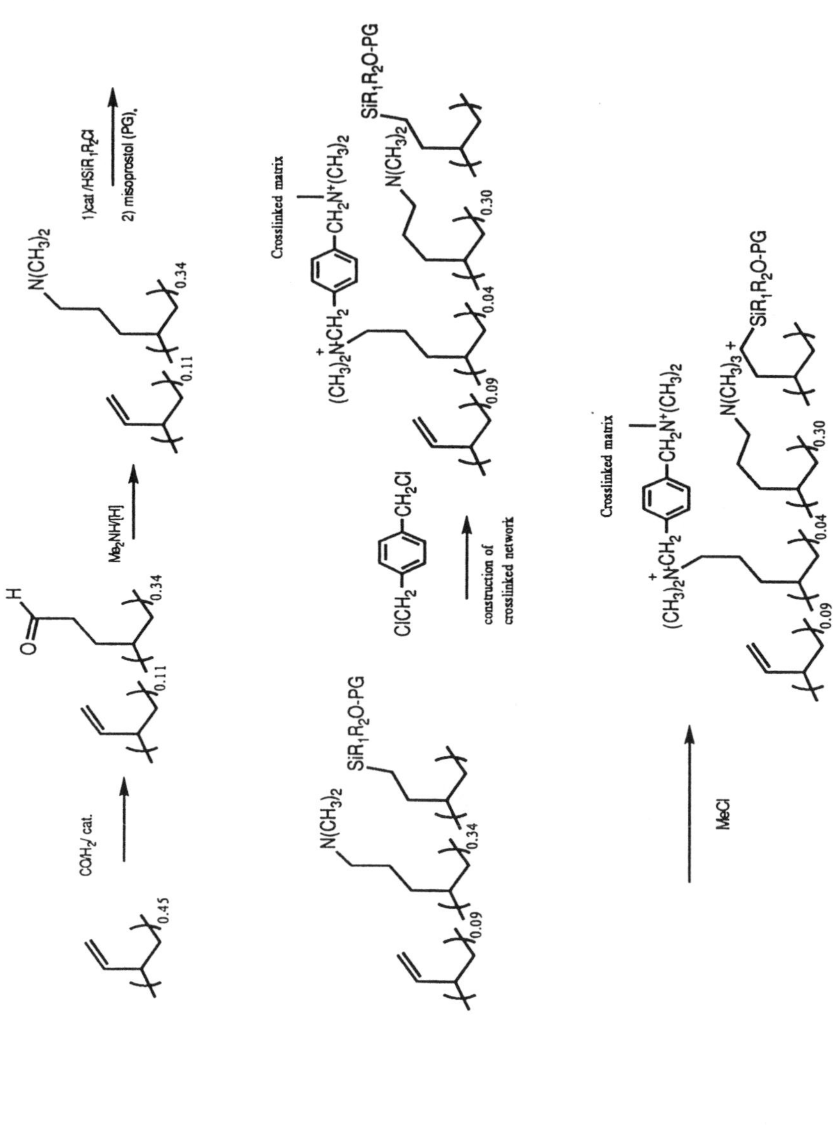

Figure 1. Synthesis of polymeric delivery systems.

readily isolated from the cyclohexane phase. The hydrosilylation of the polyamine was carried out under rigorously anhydrous conditions in a dry box. The polyamine in a toluene solution reacted with the appropriate chlorodialkyl (aryl) silane at 100°C for 24 h in the presence of tristriphenylphosphine rhodium chloride (Wilkinson's catalyst). The amounts of silane and the catalyst varied depending on the particular silane used. The chlorosilylated polymer was isolated by precipitation from DMF/THF. ^{1}H-NMR was used to determine the percentage of hydrosilylation, with a 2% incorporation customary.

Misoprostol was attached by the treatment of a chlorosilylated polyamine solution in DMF with THF solutions of imidazole and misoprostol. After about 6 h at room temperature, the excess silyl chloride groups were quenched with methanol. Since the bulky diisopropylsilyl chloride reacted very slowly under these conditions, alternative conditions were used for this system (triethylamine, dimethylaminopyridine, 50°C, 20 h). The misoprostol polymer was crosslinked using α,α-dichloro-p-xylene in THF for 60 h. The crosslinking occurs via the reaction of the dibenzylchloride with the dimethylamino groups of the polymer. The crosslinking degree in these systems was about 14% of the dimethylamino groups. The crosslinked polymer was cut with a spatula, ground in an analytical mill, and washed with THF to remove any unreacted cross-linking agent.

The final step was methylation of the remaining dimethylamino groups with methyl chloride in a Fisher-Porter pressure bottle at room temperature for 60 h. The resulting crystalline polymer was milled in a cryogenic grinder, sieved (250 microns), and finally mixed with an equal weight of hydroxypropylmethylcellulose (HPMC) to give a white, free-flowing powder. The amount of misoprostol varied among the polymer systems but generally was within the range of 3 to 5 μg misoprostol/mg material weight.

Analysis of Drug Content

A 50 mg sample was placed in a centrifuge tube. A solution of HCl (3 mL, pH 1.9) and 3 mL of HPLC MeOH were added to the sample in the tube. The mixture was stirred for certain time periods and then centrifuged for 5 min. A 250 μL sample was removed by pipette, filtered, and analyzed by HPLC. The reaction was allowed to proceed for a sufficient time period to obtain the maximum release of the drug from the matrix. The percent release at each time point was calculated using the following equation:

$$\% \text{ release at time } t \quad = \quad \frac{[\text{concentration of misoprostol}]_t}{[\text{concentration of misoprostol}]_{max}}$$

Pharmacological Assays *(5)*

Indomethacin-induced Gastric Damage. The gastric mucosal protective activity for these delivery systems was compared to that for misoprostol/HPMC *(6)* in indomethacin treated fasted rats (n = 6). Compounds were administered intragastrically immediately before each rat received 16 mg/kg of indomethacin intraperitoneally. The animals were killed 5 h later, and the glandular portion of each stomach was examined for damage. The data from two replicates for each experiment were pooled.

Diarrheagenic Activity. Evaluation for diarrheal activity was performed on 24 h fasted rats (n = 6). The animals were dosed intragastrically and then placed in individual wire mesh cages with paper lined collection trays. Diarrhea was assessed on all-or-none basis at hourly intervals for 8 h after the compound administration. Diarrhea was defined as uniformed or watery stools that wet the inner cage paper lining. The data from two replicates for each experiment were pooled, except in the evaluation of system **5.**

Results and Discussion

Chemistry

A critical component in the present series of polymer systems is the quaternary ammonium groups. Without this functionality, the polymer was extremely hydrophobic and did not release misoprostol in the aqueous acidic medium (gastric juice). The quaternary ammonium groups increased hydrophilicity of the polymer, improved its crystallinity, and caused polymer swelling in aqueous media. These groups might also provide a bioadhesive property *(7)* due to charge interaction with the mucus layer of the stomach.

The crosslinking of the polymer was performed to provide a non-absorbable matrix in order to avoid any toxicity related to the delivery system. Although this system includes a number of variables that can affect the drug release rate (*e.g.,* crosslinking density, amine concentration, *etc.*), the initial focus was on the substitution pattern of the silicon, while other variables were maintained constant. The increasing steric hindrance around the silyl ether drug bond resulted in the decreased drug release rate (Figure 2).

Pharmacology

The five delivery systems (Table I) were compared individually with misoprostol/-HPMC *(6)* for antiulcer activity against indomethacin-induced gastric damage and for diarrheagenic activity in rats. Misoprostol/HPMC was run as a control in each experiment because of the variability observed in these assays. Dimethylsilyl system **1** was the first prepared. Although the *in vitro* release rate of the drug was relatively rapid at pH 2.2 (Figure 2), the equal antiulcer activity and the reduced diarrheagenic activity of system **1** suggested that our strategy of controlled site-directed drug delivery was successful in reducing intestinal side effects while maintaining antiulcer efficacy. As a follow-up to the pharmacological data, a blood study was carried out in rats to compare the concentration of the free acid

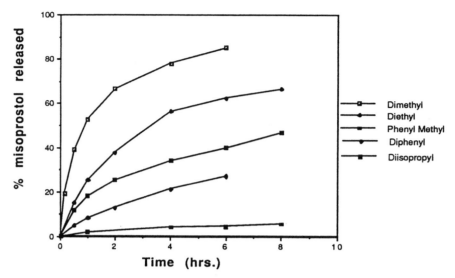

Figure 2. *In vitro* release rate of misoprostol from polymer delivery system at pH 2.2.

metabolite of misoprostol (SC-30695) after bolus intragastric doses of misoprostol/-HPMC (400 μg misoprostol/kg) and 1 (100 and 400 μg misoprostol/kg). Shown in Figure 3 is that the polymeric delivery system substantially reduced the peak plasma level. At the 400 μg/kg dose, a low level of prostaglandin was sustained during the experiment (6 h). At equal doses of misoprostol (400 μg/kg), the systemic exposure to SC-30695 (AUC) from 1 was about 58% of that for the misoprostol/HPMC. These data suggest a sustained availability of misoprostol from the delivery system in the stomach, as well as a potential for reducing systemic side effects, such as uterotonic activity.

Systems 2-5 were designed to progressively enhance the steric hindrance around the silyl ether and thus incrementally reduce the rate of drug release. Shown in Figure 2 is the progressive reduction in the drug release rate at pH 2.2. While systems 2 and 3 did not show improvement over 1 in separating the desired pharmacological effects and side effects, diphenyl system 4 was superior to 1 in terms of reduced diarrheagenic activity.

Thus, it appears that the progressive decrease in the drug release rate reduces the intestinal side effects without affecting the antiulcer activity. This conclusion was dramatically confirmed with diisopropyl system 5 which exhibited the lowest drug release rate *in vitro*, had antiulcer efficacy equal to that for misoprostol/HPMC, and displayed no evidence of diarrhea at the highest dose tested.

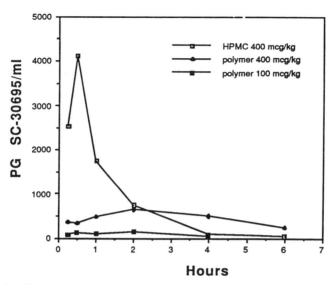

Hours

Figure 3. Plasma concentration of SC-30695 after oral administration of misoprostol/polymer.

Literature Cited

1. Collins, P.W. *Med. Res. Rev.* **1990**, 10, 149.
2. Collins, P.W. *J. Med. Chem.* **1986**, 29, 437.
3. Tremont, S.J., *et al. J. Org. Chem.*, submitted.
4. Tremont, S.J., Remsen, R.E., and Mills, P.L. *Macromolecules* **1990**, 23, 1984.
5. Perkins, W.E., Collins, P.W., Bianchi, R.G., Gasiecki, A.F., Bauer, R.F., Jones, P.H., Gaginella, T.S. *Drug Development Res.* **1991**, 23, 349.
6. The commercial dosage form contains misoprostol/HPMC, a 1:99 mixture of drug to hydroxypropylmethylcellulose.
7. Ching, H.S., Park, H., Kelly, P., Robinson, J.R. *J. Pharmaceu. Sci.* **1985**, 74, 399.

RECEIVED May 17, 1993

Chapter 17

Use of α-1,4-Polygalactosamine as a Carrier of Macromolecular Prodrug of 5-Fluorouracil

Tatsuro Ouchi, Keigo Inosaka, and Yuichi Ohya

Department of Applied Chemistry, Faculty of Engineering, Kansai University, Suita, Osaka 564, Japan

α-1,4-Polygalactosamine (PGA) and N-acetyl-α-1,4-polygalactosamine (NAPGA) have been reported to be low toxic and biodegradable polymers and to act as growth inhibitors of certain tumor cells through stimulation of the host immune system. Moreover, these polysaccharides can be expected to have affinities for hepatocytes. A water soluble macromolecular prodrug of 5-fluorouracil (5FU) with affinity for tumor cells and without side-effects was obtained as CM-NAPGA/amide/pentamethylene/ester/monomethylene/5FU conjugate. The release behavior of 5FU from this conjugate and its antitumor activity were investigated *in vitro* at 37°C in physiological saline. The release of free 5FU was observed, while no 5FU derivative was detected. The cytotoxicity of CM-NAPGA/amide-/pentamethylene/ester/monomethylene/5FU conjugate was lower than that of free 5FU against *p388D1 lymphocytic leukemia*, while the conjugate was more potent than the free 5FU against HLE hepatoma cells *in vitro*. The administration of CM-NAPGA/amide/pentamet-hylene/ester/monomethylene/5FU conjugate via intraperitoneal (i.p.) transplantation/i.p. injection in mice resulted in a significant survival effect against *p388 lymphocytic leukemia*. The conjugate in the high dose range caused neither rapid decrease of body weight of the treated mice, nor acute toxicity in the treated mice.

5-Fluorouracil (5FU) has a remarkable antitumor activity which is accompanied, however, by undesirable side-effects. α-1,4-Poly-galactosamine (PGA) is a novel basic polysaccharide purified from the culture fluid of *Paecilomyces sp. I-1* and free from acute toxicity *(1)*. PGA and N-acetyl-α-1,4-polygalactosamine (NAPGA) are non-toxic, non-immunogenetic biodegradable polymers. These polymers have been reported to act as growth inhibitors of certain tumors through stimulation of the host immune system *(2)*. Similar to galactose, galactosamine has an affinity for hepatocytes *(3)* and can be used as a targeting moiety for hepatoma cells *(4)*.

0097–6156/94/0545–0204$08.00/0

Thus, PGA and NAPGA are potential drug carriers with targetability to hepatocytes.

Previously, we synthesized PGA, its oligomer (GOS), and NAPGA linking 5FU via urea bonds. We prepared PGA/urea/hexamethylene/urea/5FU, GOS/urea/hexamethylene/urea/5FU, and NAPGA/urea/hexamethylene/urea/5FU conjugates which significantly increased the survival effects against *p388 lymphocytic leukemia* in mice by intraperitoneal (i.p.) transplantation/i.p. injection *in vivo* and exhibited strong growth inhibition effects against Meth-A fibrosarcoma in mice by subcutaneous implantation/intravenous injection *in vivo*. However, the poor solubility of these compounds in water *(5)* limited their therapeutic effect.

In order to obtain a water-soluble macromolecular prodrug of 5FU, we prepared N-carboxymethyl-N-acetyl-α-1,4-polygalactosamine (N-CM-NAPGA; Mn = 3000-10000) and O-carboxymethyl-N-acetyl-α-1,4-polygalactosamine (O-CM-NAPGA; Mn = 3000-10000) as drug carriers. The present paper is concerned with the syntheses and antitumor activities of N-CM-NAPGA/amide/pentamethylene/ester/methylene/5FU conjugate 5 and O-CM-NAPGA/amide/pentamethylene/ester/methylene/5FU conjugate **6**.

Experimental

Materials. α-1,4-Polygalactosamine (PGA) from supernatant fluid of *Paecilomyces sp. I-1* strain *(6,7)* was supplied by Higeta Shoyu Co. Ltd. The supplied PGA was treated with polygalactosaminidase isolated from *Pseudomonas sp. 881* to give low molecular weight PGA. The low molecular weight PGA obtained was purified by ultrafiltration (cut off Mw 3000 and 10000) and the degree of polymerization of the PGA was estimated to be 20 to 60.

N-CM-NAPGA and O-CM-NAPGA were prepared according to the method reported previously *(8)*. The molecular weight of N-CM-NAPGA and O-CM-NAPGA was determined by GPC (column: Tosoh TSKgel GMPW; eluent: 0.1 mmol/L acetic acid buffer; detector: RI; standard: pulluran) and found to be Mw = 16000, Mn = 10900, and Mw = 16300, Mn = 6200, respectively. The degree of substitution of carboxyl group per galactosamine unit (DCM) was estimated from the ratio of absorbance at 1737 cm-1 (COO) to that at 1647 cm-1 (CONH). It was 40 mol% for N-CM-NAPGA, and 40 and 70 mol% for O-CM-NAPGA, respectively *(9)*. N,N'-Dimethylformamide (DMF) and other organic solvents were purified by distillation. 5-Fluorouracil (5FU), 1-ethyl-3-(3-dimethylaminopropyl)-carbodiimide (EDC·HCl), and other materials were commercial grade reagents and used without further purification.

Tumor Cell Line. The *p388D1 lymphocytic leukemia* cell line was maintained in RPMI-1640 medium containing 10% heat-inactivated fetal calf serum (FCS), 2 mmol/L L-glutamine, 18 mmol/L sodium hydrogen carbonate, and 60 mg/L kanamycin. The HLE human hepatoma cell line was maintained in Dulbecco's modified Eagle minimum essential medium containing 10% FCS, 2 mmol/L L-glutamine, 18 mmol/L sodium hydrogen carbonate, and 60 mg/L kanamycin.

Synthesis of 1-[(Amino-n-pentyl)-ester]-methylene-5FU Hydrochloride 4.
1-[(Amino- n-pentyl)-ester]-methylene-5FU hydrochloride **4** was prepared through the reaction steps shown in scheme I (10). Formalin (14.4 mL of 37 vol%, or 100 mmol) was added to 17.8 g (220 mmol) of 5FU. The reaction mixture was stirred at 60°C for 45 min. The solvent was then evaporated to give the oily 1,3-dimethylol-5FU **2**. DCC (12.3 g, 120 mmol), t-butoxycarbonyl-amino-n-capronic acid (15.6 g, 120 mmol), and 0.7 g of 4-dimethylaminopyridine (DMAP) were added to the solution of **2** in 300 mL of acetonitrile and then stirred for 4 h at room temperature. The DCU formed was filtered off and washed with dichloromethane. The resulting solution was washed three times with 1 N HCl and aqueous saturated solution of NaCl. The organic layer was dried over anhydrous sodium sulfate. The solvent was evaporated to give an oily residue. The crude product was purified by recrystallization from diethyl ether to afford a white solid of 1-[(t-butoxycarbonyl-amino-n-pentyl)-ester]-methylene-5FU **3**. Compound **3** was dissolved in dioxane containing 4 N HCl, stirred for 1 h at room temperature. The solution was evaporated to give the white powder of 1-[(amino-n-pentyl)-ester]- methylene-5FU hydrochloride **4**. M.p. 188-189°C; yield 30%. IR (KBr): 3050 (N^+H_3), 3000, 2950 (CH_2), 1720 (C=O of 5FU), 1740, (COO), 1270 (C-F), 1200 (COO), 1110 cm^{-1} (C-N). ^{1}H-NMR((CD_3)$_2$SO), δ 1.34 (m,2H,CH_2), 1.6 (m,4H,CH_2), 2.35 (t,2H,CH_2), 2.73 (m,2H,CH_2), 5.58 (s,2H,CH_2), 8.05 (s,3H,N^+H_3), 8.25 (d,1H,H-6). ^{13}C-NMR ((CD_3)$_2$SO)), δ 23.9, 25.4, 26.8, 33.2, 40.3, 70.8 (CH_2), 129.7, 138.6, 140.9, 157.7 (5FU), 172.8 (C=O). UV (MeOH): λ_{max} 262nm, ε_{max} 8800.

Bonding of 5FU to N-CM-NAPGA and O-CM-NAPGA. The bonding of 5FU to N-CM-NAPGA was carried out according to Scheme II. N-CM-NAPGA (0.5 g) and 0.37 g of EDC·HCl were dissolved in 1 mL of water. The ice-cooled aqueous

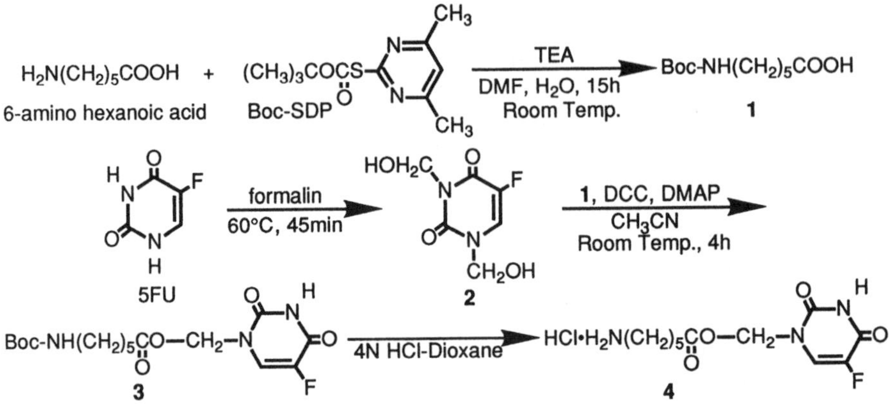

Scheme I. Preparation route for 1-[(amino-n-pentyl)-ester]-methylene-5FU hydrochloride **4**.

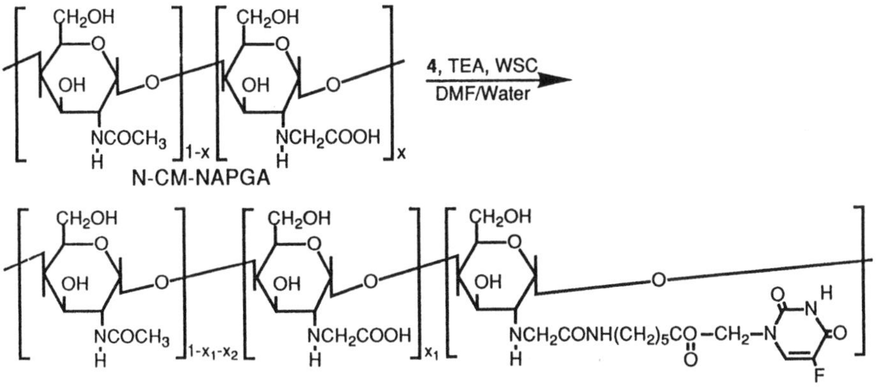

Scheme II. Synthesis of N-CM-NAPGA/amide/C$_5$/ester/C$_1$/5FU conjugate **5**.

solution was added to 1.19 g of **4** dissolved in 5 mL of DMF containing equivalent amount of triethylamine (TEA). The reaction mixture was stirred at 0°C for 1 h, and then at room temperature for 24 h. After evaporation of DMF, the crude product was dissolved in water and subjected to gel-filtration chromatography (Sephadex G-15, eluent: water). The high Mw fraction with UV absorbance characteristic of 5FU was collected and freeze dried *in vacuo* to give the resulting N-CM-NAPGA/amide/C$_5$/ester/C$_1$/5FU conjugate **5**. The bonding of 5FU to O-CM-NAPGA was carried out according to Scheme III using O-CM-NAPGA (DCM = 40, 70 mol%) to give O-CM-NAPGA/amide/C5/ester/C1/5FU conjugate **6**. The amount of 5FU (mol%) per galactosamine unit (D5FU) of the conjugates was determined by GPC of the 5FU released after the conjugate hydrolysis with 3N NaOH (column: Shodex OHpak KB-803; eluent: water; detector: UV 254.5 nm). The 5FU bonding to N-CM-NAPGA and O-CM-NAPGA was confirmed by UV absorbance due to the presence of 5FU residue in the high Mw fraction (λmax = 262 nm, in H$_2$O), and by the enhanced absorbance of amide bands (1646, 1556 cm^{-1}) in the IR spectra.

Determination of the Extent Release of 5FU from Conjugates. The release behavior of 5FU from the conjugates **5**, **6**, and PGA/urea/C6/urea/5FU was investigated in 1/15 mol/L KH$_2$PO$_4$-Na$_2$PO$_4$ buffer solution (pH = 7.4) at 37°C *in vitro*. The amount of the 5FU released from the conjugates was estimated by GPC (column: Shodex OHpak KB-803; eluent: water; detector : UV 265 nm).

Measurement of Cytotoxic Activity *in vitro.* The cytotoxic activity of the conjugates was measured against *p388D1 lymphocytic leukemia* cells and HLE human hepatoma cells *in vitro*. The tumor cell suspension (100 μL) containing 1 x 10^4 *p388D1 lymphocytic leukemia* cells in culture medium with 10% FCS was placed in a 96-wells multi-plate (Corning 25860MP) and incubated with conjugate **5**, **6**, or

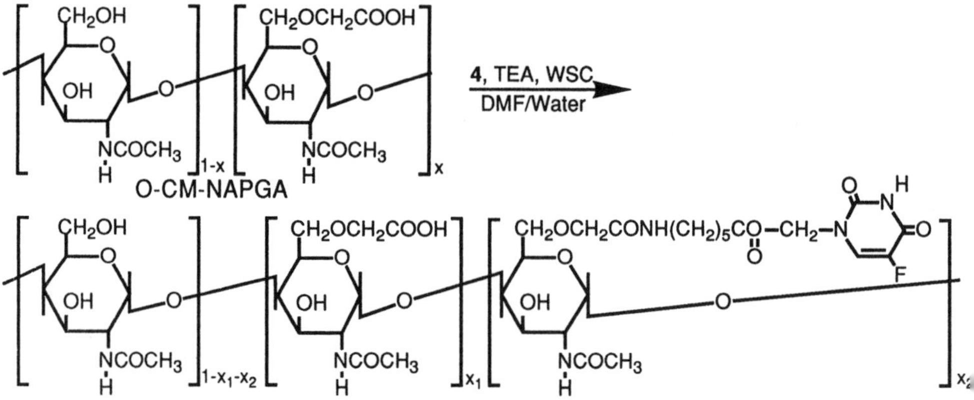

Scheme III. Synthesis of O-CM-NAPGA/amide/C_5/ester/C_1/5FU conjugate **6**.

free 5FU in a humidified atmosphere (5% CO_2) at 37°C for 48 h. The number of viable cells was determined by MTT [3-(4,5-dimethylthiazol-2-yl)-2,5-diphenyl tetrazolium bromide] assay *(11)*, using a microplate reader (MTP-120, Corona Electric Co.).

On the other hand, aliquots of tumor cell suspension (100 μL) containing 8.0 x 10^4 HLE human hepatoma cells in culture medium with 10% FCS were placed in wells of 96-well multi-plate (Corning 25860MP) and incubated in a humidified atmosphere (5% CO_2) at 37°C for 48 h. The cells, after washing with the culture medium, were added to 100 mL of the fresh culture medium containing 20 μL of PBS solution of conjugates **5**, **6**, or free 5FU, and incubated under the same condition for 48 h. The number of viable cells was determined by the method described above. The cytotoxic activity was calculated by the following equation:

$$\text{Cytotoxic activity } (\%) = (N_C - N_T)/N_C \times 100$$

where N_C is the number of control cells; N_T is the number of treated cells.

Measurement of Antitumor Activity *in vivo*. The survival effects for N-CM-NAPGA/5FU conjugate **5** and 5FU derivative **4** were tested against *p388 lymphocytic leukemia* in female CDF1 mice (30 untreated mice/group and 6 treated mice/group) *in vivo* i.p./i.p. according to the protocol of the Japanese Foundation for Cancer Research (JFCR). 1 x 10^6 of leukemia cells were injected to mice i.p. on day 0. The conjugates were dissolved in a sterile normal saline solution and administrated i.p. The mice received two doses of 200-800 mg/kg of the conjugate on day 1 and 5. The ratio of life prolongation, **T/C** (%), which is the ratio of median survival for the treated mice (**T**) to that for the control mice (**C**), was evaluated as a survival effect. The average **C** value was generally 10 days. The **T/C** (%) values over 120 were estimated as active. The *in vivo* screening was performed at the Cancer Chemotherapy Center of the Japanese Foundation for Cancer Research.

Results and Discussion

Synthesis of N-CM-NAPGA/5FU and O-CM-NAPGA/5FU Conjugates. The N-CM-NAPGA and O-CM-NAPGA reacted with 5FU derivative 4 in DMF/water solvent by using EDC·HCl as a condensation reagent to give N-CM-NAPGA amide/-C_5/ester/C_1/5FU conjugate 5, and O-CM-NAPGA/amide/C_5/ ester/C^1/5FU conjugate 6, respectively. The reaction conditions, the degree of 5FU introduction per galactosamine unit (D5FU) for conjugates, and the water solubility data are summarized in Table I.   These conjugates were easily separated from non-immobilized 4 by gel filtration chromatography. The formation of amide bond was observed in IR spectra for conjugates 5 and 6. These conjugates were water-soluble.

Release Behavior of 5FU from the Conjugate *in vitro*. The release behaviors of 5FU from N-CM-NAPGA/amide/C_5/ester/C_1/5FU conjugate 5, O-CM-NAPGA/amide/C_5/ester/C_1/5FU conjugate 6, and PGA/urea/C6/urea/5FU conjugate were investigated in 1/15 mol/L KH_2PO_4-Na_2PO_4 buffer solution (pH = 7.4) at 37°C *in vitro*.   In these release tests, only the formation of free 5FU was observed, while no 5FU derivative was detected. The release rates of 5FU from these conjugates are shown in Figure 1. The release rates of 5FU from conjugates 5 and 6 were lower than that from PGA/urea/C6/urea/5FU conjugate. These release results may be due to the fact that the hydrolysis rates for the ester and amide groups in conjugates 5 and 6 are lower than that for urea bonds in PGA/urea/C6/urea/5FU conjugate. The slow release of 5FU from the conjugates was achieved by bonding 5FU to N-CM-NAPGA or O-CM-NAPGA through pentamethylene/monomethylene spacer groups via amide/ester bonds. These conjugate can be expected to release free 5FU slowly after intravenous injection *in vivo*.

Cytotoxic Activity against Tumor Cells *in vitro*. The cytotoxic activity of the conjugates 5, 6 was investigated against *p388D1 lymphocytic leukemia* cells and HLE human hepatoma cells *in vitro*. The conjugates dose effects on the cytotoxic activity against *p388D1 lymphocytic leukemia* cells and against HLE human hepatoma cells are shown in Figure 2 and Figure 3, respectively. The IC_{50} values (concentration at which the cytotoxic activity is 50%) determined from the data in Figures 2 and 3, and the ratio of IC_{50} values of the conjugates to that of the free 5FU are summarized in Table II.   While these two conjugates did not show a higher cytotoxic activity against *p388D1 lymphocytic leukemia* cells, they showed a slightly higher cytotoxic activity against HLE human hepatoma cells *in vitro*. These results suggested affinities to HLE human hepatoma cells for N-CM-NAPGA and O-CM-NAPGA as polymer carriers. The receptor-mediated uptake of these conjugates into hepatoma cells through endocytosis was suggested.

Survival Effect. The survival effect of the N-CM-NAPGA/5FU conjugate 5 was tested against *p388 lymphocytic leukemia* in female CDF1 mice i.p./i.p.   The survival effect results for conjugate 5, 5FU derivative 4, and the free 5FU are presented in Figure 4. In the case of free 5FU, the life prolongation significantly decreased due to its toxicity in high dose ranges.   On the other hand, the life

Table I. Fixation of **4** onto N-CM-NAPGA and O-CM-NAPGA[a]

N-CM NAPGA g(mmol of CM)	O-CM NAPGA g(mmol of CM)	4 g(mmol)	TEA mmol	WSC (mmol)	unreacted CMmol%	D5FU mol%	Solubility in H_2O
0.50(0.96)	—	1.19(3.84)	3.84	1.92	21.9	18.1	O
0.32(0.61)	—	0.76(2.44)	2.44	1.22	15.0	25.0	O
—	0.53(0.90)	0.56(1.80)	1.80	1.35	31.3	8.7	O
—	0.28(0.74)	0.69(2.22)	2.22	1.11	59.4	10.6	O

[a]reacted at 0°C for 1h and at room temperature for 12h.

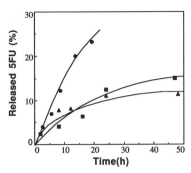

Figure 1. Release rate of 5FU from N-CM-NAPGA/amide/C_5/ester/ C_1/5FU conjugate **5**, O-CM-NAPGA/amide/C_5/ester/C_1/5FU conjugate **6**, and PGA/urea/C_6/urea conjugate in 1/15 mol/L KH_2PO_4-Na_2PO_4 buffer solution (pH = 7.4). (■) **5** (D5FU = 18.1 mol%); (▲) **6** (D5FU = 10.6 mol%); (●) PGA/urea/C_6/5FU conjugate (D5FU = 43 mol%).

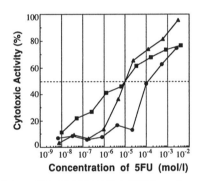

Figure 2. Cytotoxic activity of N-CM-NAPGA/5FU conjugate **5**, O-CM-NAPGA/- 5FU conjugate **6**, and free 5FU against *p388D1 lymphocytic leukemia* cells *in vitro*. (●) **5** (D5FU = 18.1 mol%); (■) **6** (D5FU = 10.6 mol%); (▲) free 5FU.

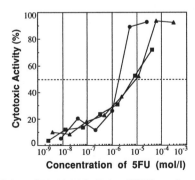

Figure 3. Cytotoxic activity of N-CM-NAPGA/5FU conjugate **5**, O-CM-NAPGA/- 5FU conjugate **6**, and free 5FU against HLE human hepatoma cells *in vitro*. (●) **5** (D5FU = 18.1 mol%); (■) **6** (D5FU = 10.6 mol%); (▲) free 5FU.

Table II.   IC$_{50}$ values of N-CM-NAPGA/5FU conjugate **5** and O-CM- NAPGA/5FU conjugate **6** against p388D1 *lymphocytic leukemia* cells and HLE human *hepatoma* cells

sample	p388D1		HLE	
	IC$_{50}$[a]	ratio[b]	IC$_{50}$	ratio
Free 5FU	1.0×10^{-5}	-	9.0×10^{-4}	-
5	1.1×10^{-4}	11.0	1.8×10^{-4}	0.20
6	1.0×10^{-5}	1.0	6.9×10^{-4}	0.76

[a]IC$_{50}$: concentration at which the cytotoxic activity shows 50%.
[b]ratio = (IC$_{50}$ of conjugate) / (IC$_{50}$ of free 5FU).

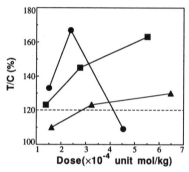

Figure 4. Survival effect per unit mol of 5FU for **4** and N-CM-NAPGA/5FU conjugate **5** against *p388 lymphocytic leukemia* in mice i.p./i.p. (●) free 5FU; (▲) **4**; (■) **5** (D5FU = 10.1 mol%).

prolongation due to conjugate **5** tended to increase with an increase in dose. While a T/C value for 5FU derivative **4** was lower than that for the free 5FU, the T/C value for conjugate **5** was as high as that for the free 5FU. Moreover, conjugate **5** did not cause a rapid decrease of body weight and did not display an acute toxicity in the treated mice even in high dose ranges, as shown in Figure 4. These results indicate that the covalent bonding of 5FU to N-CM-NAPGA effectively suppresses the side-effects of 5FU. Therefore, a novel macromolecular prodrug of 5FU was prepared by using N-CM-NAPGA as a drug carrier. The prodrug is water soluble, exhibits high antitumor activity, and reduces the side-effects typical of 5FU. Since the conjugates **5** and **6** are water soluble, they can be easily tested for antitumor activity by intravenous injection. From the standpoint of targeting, the pharmacokinetics of the conjugates with N-CM-NAPGA or O-CM-NAPGA is of interest. We will report the antitumor activity of these conjugates against hepatoma *in vivo* by intravenous injection in next paper.

Acknowledgement. The authors wish to express their sincere appreciation to Dr. Tazuko Tashiro of Cancer Chemotherapy Center of the Japanese Foundation for Cancer Research for the screening test of the survival effect against p388 lymphocytic leukemia in mice i.p./i.p. The authors wish to give their thanks to Higeta Shoyu Co. Ltd. for providing PGA.

Literature Cited

1. Takagi, H., Kadowaki, K., *Agric. Biol. Chem.* **1985**, 49, 3159.
2. Ishitani, K., Suzuki, S., Suzuki, M. *J. Pharmacobiol. Dyn.* **1988**, 11, 58.
3. Ashwell, G., Hardford, J. *Ann. Rev. Biochem.* **1982**, 51, 531.
4. Duncan, R., Hume, I. C., Kopeckova, P., Ulbrich, K., Strihalm, J., Kopecek, J. *J. Control. Rel.* **1989**, 10, 51.
5. Ohya, Y., Huang, T. Z., Ouchi, T., Hasegawa, K., Tamura, J., Kadowaki, K., Matsumoto, T., Suzuki, S., Suzuki, M. *J. Control. Rel.* **1991**, 17, 259.
6. Takagi, H., Kadowaki, K. *Agric. Biol. Chem.* **1985**, 49, 3151
7. Takagi, H., Kadowaki, K. *Chitin in Nature and Technology*, R. Muzzorelli, C. Jeuniaux, Ed. Plenum Publishing Corporation, New York, **1986**, pp. 121-128.
8. Ohya, Y., Inosaka, K., Ouchi, T., Hasegawa, K., Arai, Y., Kadowaki, K., Matsumoto, T., Suzuki, S., Suzuki, M. *Carbohydr. Polym.*, submitted.
9. Nishimura, S., Ikeuchi, Y., Tokura, S. *Carbohydr. Res.* **1984**, 134, 305.
10. Ohya, Y., Kobayashi, H., Ouchi, T. *React. Polym.* **1991**, 15, 153.
11. Mosmann, T. *J. Immunol. Methods* **1983**, 65, 55.

RECEIVED May 21, 1993

Chapter 18

Poly(methacrylic acid) Hydrogels for Rumen Bypass and the Delivery of Oral Vaccines to Ruminants

T. L. Bowersock[1], W. S. W. Shalaby[2], W. E. Blevins[1], M. Levy[1], and Kinam Park[2]

[1]School of Veterinary Medicine and [2]School of Pharmacy, Purdue University, West Lafayette, IN 47907

Poly(methacrylic acid) hydrogels were investigated for the delivery of a model antigen to the lower gastrointestinal tract of a sheep. To determine a formulation of the polymer that would deliver hydrogel without being retained in the first stomach of a ruminant, the hydrogel samples were tested by loading them with a radiopaque material (Gastrografin) and administering them orally to sheep. The hydrogel readily bypassed the first stomach and swelled in the lower gastrointestinal tract releasing Gastrografin. This hydrogel formulation was then tested to determine whether it would reach the lower small intestinal tract to release an antigen and administered to a sheep. Chromium was detected in the sheep intestine for 96 h; the peak occurred 2-15 h after administration. The results of this study indicate that poly(methacrylic acid) hydrogels could be used to administer vaccines orally to ruminants.

Bovine respiratory disease complex costs the USA cattle industry over $100 million a year (1). The available vaccines to prevent pneumonia have not been successfully used as the cattle are usually vaccinated at sale barns or at stockyards after they have already been stressed and exposed to a number of pathogens. Vaccines would be more efficient if: 1) the vaccines were given before the calves are sold to give them time to respond immunologically, i.e., when still on pasture, 2) the vaccines were given 2 times at the feedlot to induce an adequate level of immunity, or 3) they were given at the feedlot in a way that was less stressful to the calves. None of these methods are routinely utilized at this time due to the labor associated with injecting each animal individually. An efficient way to vaccinate the cattle would be through the feed or water.

Mucosal immunity is important as the first line of defense against infectious agents. Mucosal immunity prevents the attachment of pathogens to mucosal epithelium, neutralizes viruses and bacterial toxins which allows other factors of the immune system to phagocytize and remove pathogens from the mucosal site (2).

0097–6156/94/0545–0214$08.00/0

The direct application of a vaccine to the mucosal surface where microorganisms attach and replicate is the best way to induce a local immune response. However, this approach is not always practically feasible. Much handling of individual animals is required for vaccination, the antigens used may be toxic to the mucosal surface, and if modified live organisms are used, there is a danger that the vaccine can revert to a wild type and cause infection. Oral inoculation is an alternative way to stimulate a mucosal immune response. The oral administration of antigens avoids the side effects associated with the parenterally administered vaccines while stimulating the mucosal immunity at a variety of sites *(3)*.

The oral vaccination is possible due to the interaction with the mucosal immune system. When the mucosal-associated lymphoid tissue of the gut (gut-associated lymphoid tissue or GALT) or lung (bronchus-associated lymphoid tissue or BALT) is exposed to an antigen, antigen specific lymphoblasts are produced. The lymphoblasts are stimulated at the Peyer's patches, enter the lymph ducts, move through the mesenteric lymph nodes, and enter the general circulation. These lymphoblasts migrate to all other mucosal sites where they produce antibodies *(4)*. A population of memory lymphocytes also migrate to other mucosal sites, mature and produce antibodies at a later time in response to the antigen *(5)*. Studies have been previously carried out that have documented the efficacy of oral vaccines in stimulating immunity in saliva, lungs and mammary tissue in primates, mice, and cattle *(6-8)*. Recent studies performed in our laboratories have shown that it is possible to stimulate antibodies to *P. haemolytica*, the most common cause of bacterial pneumonia in cattle, following intraduodenal inoculation *(9-10)*.

Oral vaccines, however, have not been used in cattle since the administration of unprotected protein would be microbially degraded in the rumen. The rumen is the first of the four stomachs present in the cattle. In this compartment, fermentative digestion by bacteria and protozoa break down cellulose and other complex nutrients *(11)*. Therefore, an oral vaccine for cattle must bypass the rumen in order to deliver antigens to the GALT in the lower gastrointestinal tract (GIT). This would require a delivery method that protects the vaccine antigens from the degradation in the rumen. Ideally, such a delivery method should also provide a sustained release of the antigens since the optimal mucosal immunity is achieved by the repeated stimulation of the mucosal immune system *(12)*. Hydrogels have been previously used in animals for controlled drug delivery *(13)*. In this study, we suggest that hydrogels can bypass the rumen and deliver antigens in a sustained release manner to the GALT of cattle.

Materials and Methods

Hydrogel Preparation. Poly(methacrylic acid) (PMA) hydrogels were produced by cross-linking a 40 % solution of methacrylic acid (Aldrich) with a 0.8% solution of N,N'-bis-acrylamide (Bio-Rad Laboratories). Ammonium persulfate (Polysciences, Inc.) was used as the initiator and sodium bisulfite (J.T. Baker Chemical Co.) as the co-initiator. The solutions were degassed and purged with nitrogen. The monomer solutions were placed into 1 mL syringe barrels. The polymerization was carried out at 60°C for 18 h under nitrogen. The gel was removed from the syringes, cut into disks (5 mm in diameter and 3 mm thick), washed repeatedly in distilled deionized water, and dried at 37°C for 1 week.

Rumen Bypass Studies. The passage of the rumen by hydrogels was studied by administering the hydrogel loaded with a radiopaque material to a sheep. A sheep was used because it is a smaller ruminant which is easier to radiograph. The PMA hydrogel was loaded with the radiopaque marker diatrizoate meglumine/ sodium diatrizoate (Gastrografin, Squibb Diagnostics) as previously described *(13)*. The hydrogel was allowed to swell in a 45% (v/v) solution of Gastrografin for 32 h at 37°C. The hydrogel was then air dried for 1 week and oven dried at 37°C for another week. Three hundred Gastrografin-loaded hydrogel samples were administered to a sheep using a balling gun. The movement of the hydrogel through the upper GIT and the release of Gastrografin from the hydrogel was monitored over time by radiography.

Antigen Release Studies. Hydrogels as a vaccine delivery system were studied using chromium-ethylenediamine tetra-acetic acid (Cr-EDTA) as a model antigen *(14)*. Chromium-EDTA (Cr-EDTA) was chosen for this study because it is absorbed by the ruminant gastrointestinal tract in the amount of 2-3% Therefore, it can be readily detected in the intestinal fluid. Chromium-EDTA was prepared by a method previously described *(14-15)*. The dried PMA hydrogel was loaded with Cr-EDTA by swelling in a 10% (w/v) Cr-EDTA solution for 48 h at 37°C. Each hydrogel was loaded with an average of 14 mg of chromium. After loading, the hydrogels were dried at 37°C for 1 week. A total of 300 of the Cr- EDTA loaded hydrogels were placed in two 7.5 mL gelatin capsules and administered to a sheep by a balling gun.

Samples of the ileal contents were collected over 96 h through a cannula placed in the ileum 6 cm proximal to the ileal-cecal junction. The cannula was placed by transecting the ileum, suturing the cut ends, and inserting a T-shaped cannula in the proximal and distal ends *(15)*. The ends of each cannula were passed through the flank by blunt dissection and clamped securely to the skin. The two ends were connected so that all intestinal contents passed through the cannula. The intestinal contents were collected for 120 h in each study to look for hydrogels. The ileal contents were collected, poured through a 60 mesh sieve, and examined for hydrogels. The liquid portion was injected into the caudal part of the cannula. The ileal samples (50 mL) were collected, centrifuged to remove large debris, and stored at -20°C until assayed. All feces were collected for one week, broken apart, and passed through a 60 mesh sieve to look for hydrogels.

Analysis of Samples for Chromium. The intestinal samples were assayed for chromium content by atomic absorption spectrophotometry as previously described *(14,16)*. The samples of ileal contents were centrifuged at 500 g for 30 min to separate the fibrous matter from the liquid fraction. An aliquot (0.8 mL) of the supernatant was then added to 15 mL of 1 N HNO_3 and heated for 6 h at 60°C to precipitate any soluble proteins. The samples were filtered through a 0.2 um pore size filter (Sigma Chemical Company). Chromium levels were detected by atomic absorption spectrophotometry (Perkin-Elmers, Model-2380) at 357.9 nm and a slit setting of 0.7 nm. The samples containing 5 ug/mL and 15 ug/mL of chromium in 1 N HNO_3 were used as standards.

Results and Discussion

Rumen Bypass Studies. The physical arrangement of the four stomachs of a ruminant is shown in Figure 1. A significant number of PMA hydrogels entered or were about to enter the reticulum 15 min after administration. As shown in the radiograph, approximately 60% of the gel were in the reticulum within 45 min after administration (Figure 2). As the gel began swelling, Gastrografin was released due to its pH-dependent solubility *(13)*, and 3 h after administration the outline of the omasum and abomasum could be seen (Figure 3). After 3.5 h, the hydrogel could not be seen as a result of the Gastrografin release. The hydrogels appeared to remain in the reticulum; no hydrogels were seen in either the omasum or abomasum during the study.

Antigen Release Studies. The Cr-EDTA was detected in the ileum 3-96 h after administration with the peak levels at 12-15 h (Figure 4). Significant amounts of

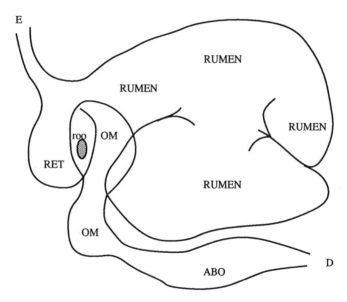

Figure 1. The four stomachs in ruminants. The ingested matter moves from the esophagus (E) into the first stomach (rumen), into the second stomach (reticulum, RET), through the reticulo-omasal orifice (roo) into the third stomach (omasum, OM), and then into the fourth stomach (abomasum, ABO) before entering the duodenum (D).

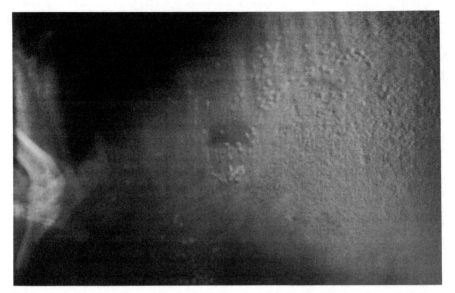

Figure 2. Lateral radiograph of the abdomen of a sheep 45 min after administration of Gastrografin-loaded hydrogel.

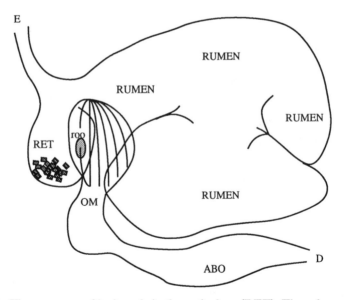

Figure 3. The presence of hydrogels in the reticulum (RET). Three hours after the hydrogel administration, the leaves of the omasum were highlighted by Gastrografin (vertical lines).

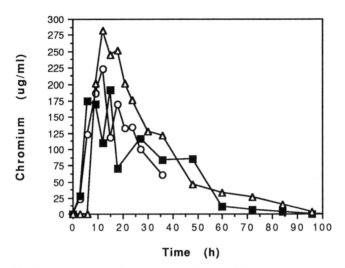

Time (h)

Figure 4. The levels of chromium detected in the ileal contents over time. The triangle, circle, and square represent results of 3 separate studies in the same sheep.

Cr-EDTA were detected up to 36 h. The Cr-EDTA levels declined rapidly between 36 and 96 h. No gel was found in the ileal contents or the feces over the course of the study.

Conclusions

This study demonstrated that hydrogels can bypass the rumen and move rapidly into the reticulum. The hydrogels' passage through the rumen depends on the size and density of the hydrogel particles. The hydrogel particles that were either too large and light, or too small and too light would be more likely to become suspended in the rumen and not enter the reticulum. This was found to be the case in our preliminary studies (data not shown). Retention time in the rumen can vary up to 54 h *(17)*. The prolonged retention in the rumen with the release of the antigens would result in microbial degradation of the antigens. Particles with a specific gravity of 1.17 to 1.77 and a diameter less than 6 mm have a shorter retention time in the rumen. Such particles move rapidly to the abomasum or the 4th stomach *(17-20)*. The PMA hydrogels used in this study entered the reticulum or the second stomach within 15 min after administration and were able to bypass the rumen efficiently due to their optimal size (5 mm in diameter, 3 mm in length) and density (specific gravity of 1.4).

Although gel retention and subsequent gel erosion appear to occur in the reticulum as a result of size discrimination at the reticulo-omasal orifice, further studies are needed to confirm this observation. Hydrogel erosion could be caused by the reticular contractions and the presence of fibrous matter in the diet that would facilitate mechanical breakdown *(22)*. This study showed that while retained in the upper gastrointestinal tract, the hydrogels released a model antigen (Cr-

EDTA) into the ileum for 96 h. The detection of Cr-EDTA in the intestinal contents was probably due to its release in the upper gastrointestinal tract rather than to the hydrogel traversing the lower intestinal tract. This is supported by the failure to find the hydrogels in the intestinal contents or feces during the studies. Moreover, the detection of Cr-EDTA 3 h after administration of the hydrogel suggests that Cr-EDTA was released promptly from the hydrogels and moved down the GIT in the liquid portion of the chyme. In summary, this study demonstrated that hydrogels are suitable to deliver vaccines orally to ruminants.

Acknowledgements. The authors would like to thank Ms. Della Ryker and Mr. Bill Reese for their excellent technical assistance. This project was supported by the Animal Health and Disease Research Funds at the School of Veterinary Medicine, Purdue University, West Lafayette, IN.

Literature Cited

1. Wohlgemouth, K.; Herrick, J.B. *Norden News* **1987**, Summer, 32-36.
2. Husband, A.J. *Prog. Vet. Microbiol. Immun.* **1985**, 1, 25-57.
3. Mestecky, J. *J. Clin. Immunol.* **1987**, 7, 265-276.
4. Rudzik, R.; Perey, D.; Bienenstock, J. *J. Immunol.* **1975**, 114, 1599-1604.
5. Elson, C. *Immunoreg.* **1985**, 8, 1-15.
6. McGhee, J.R.; Michalek, S.M. *Ann. Rev. Microbiol.* **1981**, 35, 595-638.
7. Reuman, P.D.; Keely, S.P.; Schiff, G.M. *J. Med. Virol.* **1990**, 32, 67-72.
8. Chang, C.C.; Winter, A.J.; Norcross, N. L. *Infect. Immun.* **1981**, 31, 650-659.
9. Bowersock, T.L.; Walker, R.D.; McCracken, M.D.; Hopkins, F.L.; Moore, R.N. *Can. J. Vet. Res.* **1989**, 53, 371-377.
10. Bowersock, T.L.; Walker, R.D.; Samuels, M.L.; Moore, R.N. *Can. J. Vet. Res.* **1992**, 56, 142-147.
11. Van Soest, P.J. In *Nutritional Ecology of the Ruminant*; O. and B. Books: Corvalles, Oregon, 1982; p 221.
12. Andre, C.; Bazin, H.; Heremans, J.F. *Digestion* **1973**, 9, 166-203.
13. Shalaby, W.S.W.; Blevins, W.E.; Park, K. In *Polymeric Drugs and Drug Delivery Systems*; Dunn, R.; Ottenbrite, R.M., Eds.; American Chemical Society Symposium Series No. 469; American Chemical Society: Washington, DC, 1991; pp 237-248.
14. Uden, D.; Colucci, P.; Van Soest P. *Food Agric.* **1980**, 31, 625-632.
15. Levy, M.; Merritt, A.M.; Levy, L.C. *Corn. Vet.* **1990**, 80, 143-151.
16. Kotb, A.R.; Luckey, T.D. *Nutri. Abst. Rev.* **1972**, 42, 813-844.
17. Poppi, D.; Norton, B. *J. Agric. Sci.* **1980**, 94, 275-280.
18. Kaske, M.; Hatipoglu, S.; Engelhardt W. *Vet. Scand. Suppl.* **1989**, 86, 53-54.
19. Pell, A.; Welch, J.; Wu S. *J. Control. Rel.* **1988**, 8, 39-44.
20. Welch, J. *J. Dairy Sci.* **1986**, 69, 2750-2754.
21. desBordes, C.; Welch, J. *J. Anim. Sci.* **1984**, 59, 470-475.
22. Ruckebusch, Y. In: *The Ruminant Animal, Digestive Physiology and Nutrition*; Church, D.C., Ed.; Prentice Hall: Englewood Cliffs, New Jersey, 1979; pp 65-107.

RECEIVED March 16, 1993

Chapter 19

Polymer–Solvent Interactions Studied with Computational Chemistry

Samuel J. Lee and Kinam Park

School of Pharmacy, Purdue University, West Lafayette, IN 47907

The computer simulation method which is based on a mathematical model was utilized to study the polymer-solvent interaction. The molecular dynamics was carried out for a model hydrophilic polymer poly(ethylene oxide) (PEO). The use of different dielectric constants to simulate the solvent attenuation of the interaction resulted in different chain conformations. This clearly shows the importance of the solvent in determining the structure of polymer. In a second series of calculations, molecular dynamics runs were carried-out on two three-by-three matrix of PEO chains (8 residues each chain) at dielectric constant values of 1 and 80. A third molecular dynamics run was carried out on a matrix with explicitly added water molecules. The results show that the calculation with water molecules describe a more expanded system compared to the other two calculations.

Many important phenomena occurring in the nature are determined by the interactions of solute with solvent molecules (1,2). In order to understanding these interactions, the solvent has usually been considered in a macroscopic sense. Solvent, however, should not just be considered as a macroscopic continuum characterized only by physical constants such as density, dielectric constant, or index of refraction. It should instead be considered as a discontinuum which consists of individual, mutually interacting solvent molecules. The microscopic level understanding of molecular interactions among solvent molecules and between solvent and solute can help elucidate the phenomena occurring at the macroscopic level.

Computer simulation is a useful tool in the study of properties and behavior of a variety of systems at the molecular level. Recently, the computer simulation on polymer system has been covered in a book edited by Roe (3). The molecular mechanics studies utilize a mathematical model that represents a potential energy

0097–6156/94/0545–0221$08.00/0

surface for the molecule of interest. The potential energy surface describes the energy of the system as a function of the three-dimensional structure of the molecules. Various techniques are available to study the energetics, structural, and dynamic aspects of a system. The computational resources have become sufficiently powerful and abundant to enable simulations of system consisting of 10,000 to 50,000 atoms with realistic mathematical models.

The structure and biological function of many biomolecules are affected by aqueous solvation. For this reason, theoretical models of biopolymers must include these solvent effects. The unique structural behavior of water, especially the water immediately adjacent to solutes, makes it difficult to apply continuum theories to aqueous solutions. The addition of explicit water molecules to the model has been difficult due to the dramatic increase in the number of atoms in the simulation. With the development of high speed computers, however, it is now feasible to directly model the behavior of aqueous systems through molecular dynamics and Monte Carlo simulations which specifically include solvent water molecules. Such a study using molecular dynamics has been carried out on protein dynamics in solution and in a crystalline environment by van Gunsteren and Karplus *(4)*. This computer simulation technique used in the study of biological polymers is extended to the study of the hydrophilic polymer behavior, which also depends on interactions with water.

Hydrogels play a significant role in medicine and pharmacy. The relevant background information on the hydrogels is reviewed in detail in books edited by Peppas *(5)*. Hydrogels have been used to prepare controlled release drug delivery systems. The type of drugs range from the small molecules such as flavin mononucleotide *(6)* to large molecules such as peptides and proteins *(7)*. The effectiveness of a controlled release system depends on its ability to deliver a desired amount of drug and its rate of delivery. The drug delivery rate is affected by the hydrogel's ability to interact with the solvent and subsequent ability to swell. The swelling is dependent on the hydrophilicity of the polymer, the network structure it forms, and the number of the ionizable groups in the system *(8)*. The extent of swelling controls the access of a drug with a certain size into the hydrogel network by providing the water-filled pores. The swelling of hydrogels is an important event in the process of loading of drugs, especially large molecular weight drugs, as well as their release. Therefore, the understanding of the interaction between hydrogels with water and subsequent swelling phenomena is important. As an initial step, we have examined interactions between water molecules and poly(ethylene oxide) (PEO), a hydrophilic polymer, using molecular dynamics.

Intermolecular Forces

In order to fully understand the interactions between the molecules, the intermolecular forces need to be understood. For most cases, the intermolecular interactions considered are the electrostatic and van der Waals interactions. The following coulomb's equation is utilized to model the electrostatic interactions:

$$V = \frac{q_i q_j}{D r_{ij}}$$

(1)

where V is the potential energy, q is the atomic charge, D is the dielectric constant, and r is the distance between the atoms. The van der Waals interaction is modeled through the Lennard-Jones equation,

$$V = \frac{A}{r_{ij}^{12}} - \frac{B}{r_{ij}^6}$$

(2)

where A and B are the Lennard-Jones parameters. The hydrogen bonding has been modeled through the combination of electrostatic and van der Waals or explicitly through a hydrogen bonding potential,

$$V = \frac{A'}{r_{ij}^{12}} - \frac{B'}{r_{ij}^{10}}$$

(3)

where A' and B' are the hydrogen bonding parameters.

Empirical Energy Function

The molecular mechanics has become a highly useful technique for examining structural questions that interest chemists, biologists, and biomedical researchers. The utility of the molecular mechanics approach is dependent upon the accuracy of its parameters. A major concern is the concept of parameter or constant transferability which is incorporated into the force field. The transferability of parameters from one molecule to a similar structural unit in another is a fundamental assumption in molecular mechanics. This approximation, however, works quite well as many researchers have demonstrated by innumerable reports in the literature *(3,9,10)*. The users of the molecular mechanics method, however, must be careful about extending these methodologies beyond their intended level.

Unlike quantum mechanical approaches, electrons are not explicitly included in the molecular mechanical calculations. The Born-Oppenheimer approximation, which states that the electronic and nuclear motions can be uncoupled from one another and considered separately, allows the consideration of a system in terms of the nuclear structure *(11)*. A molecule is considered to be a collection of masses that are interacting with each other through harmonic forces (which is described by a ball-and-spring model). Potential energy functions are used to describe these interactions between nuclei (see below). It has become possible to use increasingly more sophisticated equations to reproduce molecular behavior as our understanding on the molecular behavior has been increased.

The potential energy of a system is based on the coordinates of its constituent atoms. The potential energy is derived using the empirical energy function *(12,13)*.

The function consists of atomic motions of the constituent atoms and the nonbonded interactions between the constituent atoms. The harmonic motion equation is used to simulate the bond stretching and the changes in the bond angles and the torsional angles. The non-bonded interaction is simulated through the use of the van der Waals equation to calculate the dispersion interactions and the use of Coulombic interaction equations to calculate the charged interactions.

With optimized parameterization, the electronic system is implicitly taken into account. The force constants have been obtained by fitting to vibrational data in some cases and from the literature for others. For the most part, geometric constants have been derived from the crystallographic data. The empirical method is based on the use of parameters which are derived from the experimental structure. Methods such as neutron diffraction and x-ray crystallography are used to obtain the experimental structure. Using this structure as the standard, one can use the empirical equation with certain parameters to fit the structure. The resulting parameters which give the best fit are used to study the physical process of the similar systems.

Molecular Dynamics

The principle of molecular dynamics is the numerical integration of the classical equations of motion for a system of interacting particles over a certain period of time. This is accomplished by using the Verlet algorithm which tracks the position of each atom at succeeding time steps *(14)*. The simulation proceeds in a series of small time increments and, after each time step, the force on each atom is evaluated. The total energy (sum of kinetic and potential energies) is calculated from the following equation:

$$E = \frac{1}{2} \sum_{i=1}^{N_{df}} m_i v_i^2 + V(r) \tag{4}$$

where m is the mass of atom, v is the velocity of atom, r is the coordinate of the atoms, and the potential energy function V is,

$$V(r) = \frac{1}{2} \sum_{bonds} K_b (b - b_o)^2 + \frac{1}{2} \sum_{\substack{bond \\ angles}} K_\theta (\theta - \theta_o)^2 +$$

$$\frac{1}{2} \sum_{\substack{torsion \\ angles}} K_\phi [1 + \cos (n\phi - \delta)] +$$

$$\sum_{\substack{nb\ pairs, \\ r<8\text{Å}}} \left[\frac{A}{r^{12}} - \frac{C}{r^6} + \frac{q_1 q_2}{Dr} \right] + \sum_{H\ bonds} \left[\frac{A'}{r^{12}} - \frac{C'}{r^{10}} \right] + \tag{5}$$

$$\sum_{\substack{\text{int}-\text{ext}\\ \text{pairs, }r<8\text{Å}}} \left[\frac{A}{r^{12}} - \frac{C}{r^{6}} + \frac{q_1 q_2}{Dr} \right] + \sum_{\substack{\text{int}-\text{ext}\\ \text{H bonds}}} \left[\frac{A'}{r^{12}} - \frac{C'}{r^{10}} \right]$$

where **b**, **Θ**, and **φ** signify bond distance, bond angle, and torsional angle, respectively, **K**'s are the corresponding constants *(15)*. The results of the comparison between different molecular dynamics algorithms show that the Verlet algorithm is more desirable than the Gear predictor-corrector algorithm, especially with the use of longer time step *(16)*.

Simulation of Water

The importance of water and aqueous solutions in many biological systems has attracted a great deal of interest. This is mainly due to the anomalous characteristics of water molecules. The unique character of water stems from the strong and directional hydrogen bonds among water molecules. The effect of the solvent is incorporated into the simulation either implicitly through a dielectric constant or explicitly through addition of solvent molecules.

The molecular dynamics and statistical mechanics simulation have been used to study the water and its characteristics. Pair-additive interactions have been assumed in computer simulations on liquid water and aqueous solutions. The pair-additive means that the energy of a given assembly is expressed as a sum of interaction energies between all pairs of molecules in the assembly. The interaction between an isolated pair of molecules is assumed to be unaffected by the presence of a third molecule. The shortfalls of the pair-additive potentials given by Barnes *et al. (17)* are: i) isolated water molecules are poor hydrogen donors and acceptors implying that its acid/base properties must be related to its bulk phase rather than to the isolated molecule; ii) quantum mechanical calculations on water trimers and tetramers, for example, suggest that hydrogen bonding in these clusters is stronger than that in the dimer by 20-30%; and iii) the dipole moment of hexagonal ice is between 2.6 D and 3.0 D while that for the isolated water molecule is only 1.85 D. To model the cooperativity of water molecules, a polarizable electropole water model was developed *(17)*. Other complex water potentials are available *(18-20)*. The success of simulation using a certain water potential is dependent on the right choice of intermolecular potential functions for the water dimer, because the parameterization for the liquid water model is based on the dimer.

Transferable Intermolecular Potential 3-Point (TIP3P) Water

The TIP3P water is a 3-point model *(18,20)*. The 3 points are represented by an oxygen atom and two hydrogen atoms of a water molecule. Two negative charges are assigned to the oxygen and a positive charge is assigned to each hydrogen atom to create three point charges. The Monte Carlo simulation of the TIP3P molecules showed the results which were very favorable with the experimental values obtained for the bulk water. The TIP3P model, however, resulted in structure which was more organized than the experimental observation beyond the first water shell. The utility of the TIP3P water potential is that the calculation of a three site model is much simpler and the computational time is significantly shorter than more complex 4- or 5-site models. A good correlation of the three site model with experimental data and the higher computational cost for more complex 4- or 5-site models argues for the use of TIP3P water model.

Experimental

The molecular dynamics and the molecular modeling were carried out on a SiliconGraphics Personal Iris 4D/30TG+ *(21)* with Quanta/CHARMm software package *(22)*.

Poly(ethylene oxide) (PEO) chains were generated using the polymer generator option on the Quanta software. The all atom topology which includes explicit hydrogen atoms was used to generate the PEO chains. The monomer topology files contain the information on the atomic coordinates and other relevant atomic informations. The 15 residue-chain was created by linking the monomers together. The energy of the resulting structure was minimized using the steepest descents minimization procedure which relieves any energetic strain. After the minimization of the structures, we carried out a 14 picosecond (ps) molecular dynamics simulation which consisted of 2.04 ps heating, 2 ps equilibration, and 10 ps simulation. The heating step brought the system temperature up to 300 K from 0 K. The time step used was 0.001 ps, and thus, each ps consists of 1000 different calculations. The resulting structures were saved at every 10 steps so in total 1000 different conformation were saved for 10 ps of simulation. In the molecular dynamics, the SHAKE algorithm *(21)* is frequently used to reduced the number of internal degrees of freedom. The SHAKE procedure applies additional forces to keep the constrained interaction at the equilibrium value. We applied the SHAKE algorithm on the hydrogen bond length to help reduce the total calculational time. The cutoff distance of 8 Å was used in the calculation of nonbonded interactions. The distance-dependent dielectric function was used in the calculation. The molecular dynamics was carried out using dielectric values of 1 and 80 to simulate the solvent attenuation of the interactions.

In the second set of the calculation, a three-by-three matrix of 8 residue-PEO chains was generated. This polymer matrix was put through the same 14 ps molecular dynamics with dielectric constant values of 1 and 80. In addition, a three-by-three polymer matrix was put into a 30 Å water box which consists of 1000 TIP3P water molecules and the 14 ps molecular dynamics was carried out.

The comparison between the hydrophilic PEO and the hydrophobic polyethylene (PE) matrices was also carried out using a longer 25 ps molecular dynamics simulations. The PE matrix was created using the same procedure as used for the PEO matrix with the polymer generator. The resulting structure was energy minimized and placed into a 30 Å water box. The systems utilized the periodic boundary conditions to simulate the bulk water around the polymer matrix. The cutoff distance utilized for the image was 9 Å. A 5 ps heating and 20 ps equilibration was carried out to obtain the equilibrated structures of the polymer chains and surrounding water molecules.

Results and Discussion

The PEO was chosen as a model polymer, because it has limited structural possibilities. Since the initial head-to-head linked straight polymer chain consisted of a relatively small energetic strains, the energy minimization occurred quite rapidly.

The use of different dielectric constants to simulate the effect of the solvent resulted in different polymer chain conformations. The experimental results from the literature of linear chain PEO showed two distinct polymer conformation in water and in nonpolar solvents. The PEO polymer chain had a helical structure in water while it had a random coil structure in nonpolar solvents *(24,25)*. The simulation of a PEO chain with dielectric constant value of 80 resulted in a structure with some helical characteristics which was more compact than the PEO chain with dielectric constant value of 1. The helical characteristics is much more evident when the ribbon structure was generated. The simulation of the polymer chain with dielectric constant value of 1 resulted in a structure which was less compact than the polymer chain with dielectric constant value of 80. This might have occurred due to the stronger repulsive interaction between ether oxygen atoms at lower dielectric constants. Smaller dielectric values lead to stronger interactions, either attraction or repulsion. Thus, the repulsion between the oxygen atoms is much stronger at dielectric constant of 1 and as a result the structure became less compact.

The 3x3 matrix of the 8 residue PEO was also subjected to the molecular dynamics simulation. The 3x3 matrix and the 8 residue length was chosen to fit the entire structure in the 30 Å water box and to provide enough water molecules to surround the polymer matrix. The interaction occurred both intermolecularly as well as intramolecularly. The use of dielectric constant value of 80 resulted in more compact system than the system simulated with dielectric constant of 1. The absence of water molecules, however, led to two immediately obvious inadequacies in the simulation. First, the hydrogen bondings were absent due to lack of water molecules even though PEO was hydrophilic. Secondly, due to the absence of water molecules surrounding the polymers, the polymer chains readily moved away from each other in a short time period *(26)*.

The result of simulations with explicit water molecules showed a more realistic representation of the interactions involved. The addition of water molecules allowed the hydrogen bondings to be readily observed between the polymer chains and the water molecules. In addition, the water molecules around the polymer matrix acted to prevent the spurious spreading of the polymer chains. Figures 1 and 2 show the configurations of PEO and polyethylene (PE) matrix after the 25 ps simulation in the presence of explicitly added water molecules. In Figure 1, the favorable interaction between PEO chains and water molecules allow movement of PEO chains into the bulk water. The spreading of the PEO chains and the more

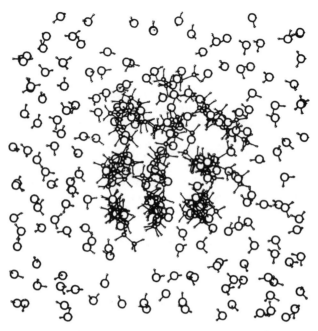

Figure 1. Representation of PEO matrix immersed in a 30 Å water box after the molecular dynamics simulation. Around 75 % of water molecules were deleted for clarity.

randomness of the PEO chains configuration indicate the favorable interaction of PEO chains with the water molecules. The extensive hydrogen bonding occurred between PEO chains' ether oxygen and water molecules. Some water molecules surrounding the PEO matrix was oriented to interact through hydrogen bond with PEO ether oxygens. The interaction of PEO chains are highly favorable with water molecules. In contrast, the interaction between water molecules and the PE chains are not favorable. Figure 2 shows the significant order in the PE matrix. The hydrogen bonding does not occur between PE chains and the water molecules due to the lack of hydrogen bond donor or acceptor atoms in the PE chains. Thus, the water molecules are oriented in such a way that the hydrogen bonding is maximized among water molecules. Figures 3 and 4 show the PEO matrix configurations before and after the simulation. As shown in Figure 3, the PEO chains are highly ordered. After the simulation, the favorable interactions with the water molecules allow the chains to move and to take on more random configuration as shown in Figure 4. In contrast, the PE chains are kept together by the water molecules around the polymer matrix. The hydrophobic interaction results from the lack of hydrogen bonding capability of PE chains and the preference of water molecules for interacting with each other instead of with PE chains. The energy

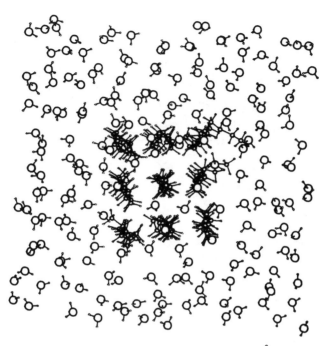

Figure 2. Representation of PE matrix immersed in a 30 Å water box after the molecular dynamics simulation. Around 75 % of water molecules were deleted for clarity. (A few water molecules which seem to be inside the PE matrix are actually in front of the polymer matrix.)

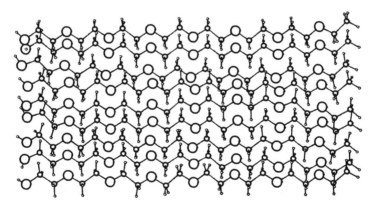

Figure 3. The configuration of PEO matrix after the initial minimization in a 30 Å water box. The water molecules are not shown.

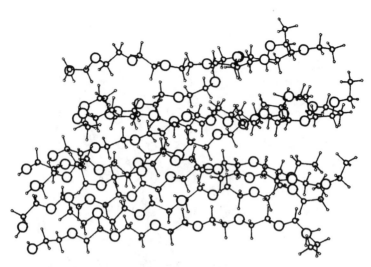

Figure 4. The configuration of PEO matrix after the molecular dynamics simulation in a 30 Å water box. The water molecules are not shown.

minimized structure of the PE matrix shown in Figure 5 is not significantly different than the minimized PEO matrix of Figure 3. The PE chains move due to the kinetic energy received through the increase in temperature. However, the hydrophobic effect imparted by the water on the PE chains keeps them together as a group even with kinetic motion of the chains. Figure 6 clearly shows the compact and more ordered configuration for the PE matrix in contrast to the PEO matrix shown in Figure 4. The two representative polymers PEO and PE, have simple structures. Even though the structure of PEO and PE differ by an ether oxygen, the interactions of these polymer chains with water molecules are quite different as shown in the simulations. The simulation of interactions of polymer chains containing bulky side groups, such as poly(hydroxyethyl methacrylate), with the water molecules is much more complex but the information acquired will highly useful.

The unique properties of water, such as the small size, highly variable orientations, and the ability to form directional interactions, make the simulation utilizing the macroscopic representation such as dielectric constants inadequate. To understand the occurrence of polymer-solvent interactions at the molecular level, the explicit addition of water molecules may be necessary to more accurately represent the system of interest.

The present simulation is modest. Much more extensive system, however, can also be simulated and for much longer simulation time period. The correctness of the result depends on the soundness of the theory which it is based upon. The fine-tuning of the computational methods is ongoing. The result of the simulations will become better and more reliable. The simplicity in building polymers and the

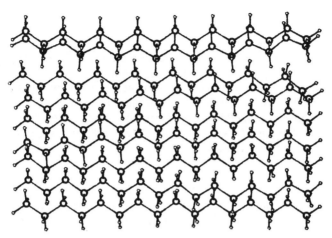

Figure 5. The configuration of PE matrix after the initial minimization in a 30 Å water box. The water molecules are not shown.

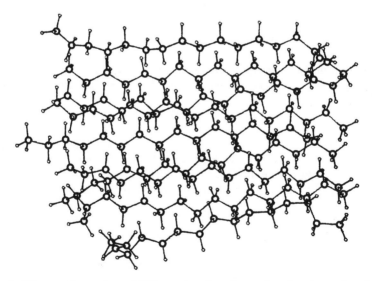

Figure 6. The configuration of PE matrix after the molecular dynamics simulation in a 30 Å water box. The water molecules are not shown.

availability of reasonable parameters make it possible to investigate the interactions of various polymers with water molecules. This simulation is currently used in our laboratory to examine the swelling properties of hydrogels.

Conclusion

The preliminary evaluation of solvent effect on the hydrophilic polymer with the computational technique indicates that the method is useful. The simulation showed the effect of water on the conformation of the hydrophilic and hydrophobic polymer chains. The use of the dielectric constant may not be able to adequately simulate the proper solvent effects. Our simulation shows that the explicit inclusion of the water molecule is necessary to more clearly understand the polymer-water interaction at the molecular level.

Acknowledgment

This study was supported by NIH through grants HL39081 and GM8298, and the Petroleum Research Fund administered by the American Chemical Society.

Literature Cited

1. Dill, K. A. *Biochemistry* **1990**, 29, 7133-7155.
2. Arakawa, T.; Kita, Y.; Carpenter, J. F. *Pharm. Res.* **1991**, 8, 285-291.
3. Roe, R. J., Ed.; *Computer Simulation of Polymers*; Prentice Hall: Englewood Cliffs, 1991.
4. van Gunsteren, W. F.; Karplus, M. *Biochemistry* **1982**, 21, 2259-2274.
5. Peppas, N. A., Ed. *Hydrogels in Medicine and Pharmacy*; CRC Press: Boca Raton, 1987; Vol I-III.
6. Shalaby, W. S. W.; Blevins, W. E.; Park, K. *J. Controlled Release* **1992**, 19, 131-144.
7. Albin, G.; Horbett, T. A.; Ratner, R. A. *J. Controlled Release* **1985**, 2: 153.
8. Gehrke, S. H.; Lee, P. I. In *Specialized Drug Delivery Systems: Manufacturing and Production Technology*; Tyle, P., Ed.; Marcel Dekker, Inc.: New York, 1990; Ch. 8.
9. Allan, M. P.; Tildesley, D. J. *Computer Simulation of Liquids*; Clarendon: Oxford, 1987.
10. French, A. D.; Brady, J. W., Ed. In *Computer Modeling of Carbohydrate Molecules*; ACS Symposium Series No. 430; American Chemical Society: Washington, DC, 1990.
11. Born, M.; Oppenheimer, R. *Ann. Phys.* **1927**, 84, 457-484.
12. Weiner, P. K.; Kollman, P. A. *J. Comput. Chem.* **1981**, 2, 287-303.
13. Brooks, B. R.; Bruccoleri, R. E.; Olafson, B. D.; States, D. J.; Swaminathan, S.; Karplus, M. *J. Comput. Chem.* **1983**, 4, 187-217.
14. Verlet, L. *Phys. Rev.* **1967**, 159, 98-103.
15. Gelin, B. R.; Karplus, M. *Biochemistry* **1979**, 18, 1256-1268.

16. van Gunsteren, W. F.; Berendsen, H. J. C. *Mol. Phys.* **1977**, 34, 1311-1327.
17. Barnes, P.; Finney, J. L.; Nicholas, J. D.; and Quinn, J. E. *Nature* **1979**, 282, 459-464.
18. Jorgensen, W. L. *J. Am. Chem. Soc.* **1981**, 103, 335-340.
19. Berendsen, H. J. C.; Grigera, J. R.; Straatsma, T. P. *J. Phys. Chem.* **1987**, 91, 6269-6271.
20. Jorgensen, W. L.; Chandrasekhar, J.; Madura, J. D.; Impey, R. W.; Klein, M. L. *J. Chem. Phys.* **1983**, 79, 926-935.
21. SiliconGraphics, 2011 N. Shoreline Blvd., Mountain View, CA 94043.
22. Quanta 3.2.3/CHARMm 21.3, Polygen Corp., 200 Fifth Ave., Waltham, MA 02254.
23. Ryckaert, J. P.; Cicotti, G.; Berendsen, H. C. *J. Comput. Phys.* **1977**, 23, 327-341.
24. Bailey, F. E., Jr.; Lundberg, R. D.; Callard, R. W. *J. Polym. Sci. Part A* **1964**, 2, 845-851.
25. Maron, S. H.; Filisko, F. E. *J. Macromol. Sci. Phys.* **1972**, 6, 79-90.
26. Lee, S, J.; Park, K. *Polymer Preprints* **1992**, 33(2), 74-75.

RECEIVED August 15, 1993

INDEXES

Author Index

Affiliation Index

Subject Index

Bestsellers from ACS Books

The ACS Style Guide: A Manual for Authors and Editors
Edited by Janet S. Dodd
264 pp; clothbound ISBN 0–8412–0917–0; paperback ISBN 0–8412–0943–X

The Basics of Technical Communicating
By B. Edward Cain
ACS Professional Reference Book; 198 pp;
clothbound ISBN 0–8412–1451–4; paperback ISBN 0–8412–1452–2

Chemical Activities (student and teacher editions)
By Christie L. Borgford and Lee R. Summerlin
330 pp; spiralbound ISBN 0–8412–1417–4; teacher ed. ISBN 0–8412–1416–6

Chemical Demonstrations: A Sourcebook for Teachers,
Volumes 1 and 2, Second Edition
Volume 1 by Lee R. Summerlin and James L. Ealy, Jr.;
Vol. 1, 198 pp; spiralbound ISBN 0–8412–1481–6;
Volume 2 by Lee R. Summerlin, Christie L. Borgford, and Julie B. Ealy
Vol. 2, 234 pp; spiralbound ISBN 0–8412–1535–9

Chemistry and Crime: From Sherlock Holmes to Today's Courtroom
Edited by Samuel M. Gerber
135 pp; clothbound ISBN 0–8412–0784–4; paperback ISBN 0–8412–0785–2

Writing the Laboratory Notebook
By Howard M. Kanare
145 pp; clothbound ISBN 0–8412–0906–5; paperback ISBN 0–8412–0933–2

Developing a Chemical Hygiene Plan
By Jay A. Young, Warren K. Kingsley, and George H. Wahl, Jr.
paperback ISBN 0–8412–1876–5

Introduction to Microwave Sample Preparation: Theory and Practice
Edited by H. M. Kingston and Lois B. Jassie
263 pp; clothbound ISBN 0–8412–1450–6

Principles of Environmental Sampling
Edited by Lawrence H. Keith
ACS Professional Reference Book; 458 pp;
clothbound ISBN 0–8412–1173–6; paperback ISBN 0–8412–1437–9

Biotechnology and Materials Science: Chemistry for the Future
Edited by Mary L. Good (Jacqueline K. Barton, Associate Editor)
135 pp; clothbound ISBN 0–8412–1472–7; paperback ISBN 0–8412–1473–5

For further information and a free catalog of ACS books, contact:
American Chemical Society
Distribution Office, Department 225
1155 16th Street, NW, Washington, DC 20036
Telephone 800–227–5558